Gauhar Syzdykova
B. K. Orazalina
G. M. Nurtasina

Environmental impacts on human health

Gauhar Syzdykova
B. K. Orazalina
G. M. Nurtasina

Environmental impacts on human health

ScienciaScripts

Imprint

Any brand names and product names mentioned in this book are subject to trademark, brand or patent protection and are trademarks or registered trademarks of their respective holders. The use of brand names, product names, common names, trade names, product descriptions etc. even without a particular marking in this work is in no way to be construed to mean that such names may be regarded as unrestricted in respect of trademark and brand protection legislation and could thus be used by anyone.

Cover image: www.ingimage.com

This book is a translation from the original published under ISBN 978-3-659-90753-1.

Publisher:
Sciencia Scripts
is a trademark of
Dodo Books Indian Ocean Ltd. and OmniScriptum S.R.L publishing group

120 High Road, East Finchley, London, N2 9ED, United Kingdom
Str. Armeneasca 28/1, office 1, Chisinau MD-2012, Republic of Moldova, Europe
Managing Directors: Ieva Konstantinova, Victoria Ursu
info@omniscriptum.com

Printed at: see last page
ISBN: 978-620-2-75278-7

CONTENTS.

INTRODUCTION

Relevance of the research topic. Modern anthropogenic factors, representing a huge variety of harmful effects on the environment, have a pronounced impact on the formation of public health; spread direct and indirect, combined and complex action of chemical, physical and biological factors.

According to WHO, the contribution of various unfavourable environmental factors to the formation of public health is 25-35%. The problem of unfavourable influence of environmental factors on health is becoming more and more urgent every year. Scientific problems of assessing the impact of environmental factors on human health and justification of state health-improving measures are today priority tasks of state policy in practically all economically developed countries.

Assessment of the significance of environmental pollution on biological responses of the human organism, on health indicators is an important component of the study of the impact of environmental factors on human health along with comparison of concentrations of individual pollutants with hygienic standards, because it integrally takes into account the impact of all, including unidentified, pollutants, their complex and combined effect on the human body (BushtuevaK.A., Sluchanko I.S., 1979).

Usually for each factor are set their own standards. Meanwhile, the combined effect of harmful factors depending on their physicochemical properties, doses (concentrations), duration, multiplicity and sequence of exposure can manifest itself in the form of additivity (the effect is equal to the sum). The effect of the environment can be the cause not only of various kinds of somatic diseases, including malignant ones, but also of genetic disorders that can manifest themselves in future generations. The problem is thus not only of medical, but also of economic and social importance. It is extremely complex and remains insufficiently researched. As a result, environmental changes lead to the emergence and increased morbidity of the population with ecodependent and ecobounded pathology (ecopathology) (Dubovoy I.I., 2007).

Ecologically conditioned are such diseases, the frequency of which is increasing against the background of increasing intensity of harmful factors, presumably influencing this type of diseases (Maimulov V.G., Nagorny S.V., Shabrov A.V., 2000).

Radioactive and chemical substances polluting atmospheric air occupy a special place among harmful factors. When assessing harmful factors, the combined or combined nature of their effect on humans is not always taken into account. Joint action of radioactive and chemical pollution of the environment creates complex hygienic problems, which are insufficiently solved, which determined the goals and objectives of this work.

Purpose of the study: integrated ecological and hygienic assessment of environmental factors and their impact on public health.

The following **research objectives** have been selected to achieve the objective of **the study:**

1) to assess complex environmental pollution of districts of Akmola region by ecological and hygienic indicators of atmospheric air, drinking water and soil;

2) to reveal the influence of a complex of environmental factors on the peculiarities of health formation in children, adolescents and adults by morbidity indicators;

3) establish a cause-and-effect relationship between environmental factors and health indicators;

4) based on the results of the study, to give recommendations to reduce the impact of anthropogenic load on the health of the population.

The object of the study is three districts of Akmola region (Akkol, Burabay, Zerendinsky).

The subject of the study is the quality parameters of environmental factors (atmospheric air, soil, water) and health indicators of children (1113 years old), adolescents (14-17 years old) and adults (18-30 years old).

The degree of study of the research topic. Studies conducted in the Republic of Kazakhstan (P.P. Petrov, T.K. Kalzhekov, 1990; N. Zhakashov, 1993; U.I. Kenesariev. 1993 and others), the dependence of negative trends in the health of the population on the complex impact of environmental and social factors, which act differently in each region has been established.

Research methods. A set of modern socio-ecological, physico-chemical, hygienic, statistical methods of research was used in the work.

Scientific novelty of the study.

Complex ecological assessment of some districts of Akmola region was carried out, priority pollutants and degree of pollution were determined.

The correlation between the degree of pollution of the oblast districts and morbidity of child, adolescent and adult population has been revealed.

For the first time at the regional level, causal relationships between environmental pollution indicators and morbidity have been established.

Recommendations to reduce the impact of anthropotechnogenic load on public health are given.

Practical significance of the work lies in the fact that in the process of the study identified the territories of Akmola region with different levels of ecological ill-being of the environment, priority pollutants and their impact on health, which makes it possible to use the obtained data in the development of preventive measures aimed at reducing the impact of anthropotechnogenic load on the health of population.

The research results were used in recommendations aimed at reducing the impact of anthropotechnogenic load on public health.

CHAPTER 1

CURRENT ASSESSMENT OF THE IMPACT OF ANTHROPOGENIC ENVIRONMENTAL POLLUTION ON PUBLIC HEALTH

1.1 Ecological characterisation of sources of environmental pollution by industrial emissions and motor vehicles

The widespread pollution of the environment by a variety of substances, sometimes completely alien to the normal existence of the human body, poses a serious threat to our health and the well-being of future generations. Therefore, environmental problems need to be addressed immediately. It is necessary to limit the detrimental impact of economic activity on the environment, to minimise emissions of harmful substances into the atmosphere [1].

Atmospheric air pollution means an increase in concentrations of physical, chemical, biological components above the level that takes natural systems out of balance. The highest concentrations of harmful substances are in the atmospheric air, which exceed the maximum permissible concentrations by 2-5 times, and it is in these areas that their main mass is accumulated on the soil and surface of water bodies. Various negative changes in the Earth's atmosphere are mainly associated with changes in the concentrations of minor components of atmospheric air [2].

The main anthropogenic sources of pollution include enterprises of fuel and energy complex, transport, various machine-building enterprises, heavy industry enterprises. The most significant of them are:

- Thermal power plants pollute the atmosphere with emissions that contain sulphur dioxide, sulphur dioxide, nitrogen oxides, soot, dust and ash that contain heavy metal salts.

- Combines of ferrous metallurgy, which include blast furnace, steelmaking, rolling mills, sintering plants, coke plants, etc.
- Non-ferrous metallurgy, which pollutes the atmosphere with non-ferrous and heavy

metal compounds, mercury vapour, sulphur dioxide, nitrogen oxides, carbonic anhydride, etc.

- Mechanical engineering and metal processing. Emissions from these enterprises contain aerosols of non-ferrous and heavy metal compounds, including mercury vapour. The oil refining and petrochemical industry is a source of such atmospheric pollutants as hydrogen sulphide, sulphur dioxide, carbon monoxide, ammonia, hydrocarbon and benzaperene.

- Organic chemistry enterprises. Emissions of large amounts of organic substances that have a complex chemical composition, hydrochloric acid heavy metal compounds, contain soot and dust.

- Enterprises of inorganic chemistry. Air emissions from these enterprises contain sulphur and nitrogen oxides, phosphorus compounds, free chlorine, and hydrogen sulphide.

- Motor transport. The geographical patterns of distribution of pollutants from it are very complex and are determined not only by the configuration of the motorway network and the intensity of motor vehicles, but also by the large number of intersections where vehicles stand for a certain period of time with their engines running. The number of vehicles worldwide is 630 million [3].

Environmental pollution from motor vehicles is one of the most unsafe for human health, because exhaust gases enter the atmosphere, where it is difficult to disperse them. Exhaust gases from motor vehicles contain large amounts of nitrogen oxide, unabsorbed carbon, aldehydes and soot, as well as carbon monoxide [4].

Due to the huge number of motor vehicles, they have a huge impact on the atmosphere and human health. It is believed that thousands of people die every year due to exhaust gases, and the damage they cause to the environment is estimated in billions of dollars. Exhaust emissions influence the development of many diseases. Industrial emissions have a negative impact on human health, destroy materials and equipment, and reduce the productivity of forestry and agriculture [5].

Nowadays, scientists are actively working on the creation of technologies for the utilisation of emissions, environmentally friendly production, fuel. Technologies for utilisation of emissions have been created. To purify emissions it is necessary to build purification facilities. If all chemical enterprises collected production emissions, they would receive tens of thousands of tonnes of such valuable substances as nitric and sulphuric acid, sulphuric anhydride, fluorine and others.

Unfortunately, the created effective production technologies are not applied at the majority of enterprises because of their high cost, and sometimes because of neglecting the environmental problem [3].

Air pollutant emissions are characterised by four attributes: aggregate state, chemical composition, particle size and mass flow rate of the emitted substance. Pollutants are emitted into the atmosphere in the form of dust, smoke, mist, vapour and gaseous substances. The most common pollutants entering the atmospheric air from anthropogenic sources are: carbon monoxide, sulphur dioxide, nitrogen oxides, hydrocarbons, dust, carbon monoxide - the most common and most significant atmospheric impurity, called carbon monoxide in everyday life. The content of CO in natural conditions is from 0.01 to 0.2 mg\m^3, but in large cities its content ranges from 1-210 mg\m^3. The highest concentration is observed in streets and squares of cities with heavy traffic, especially at intersections. Its specific weight is more than 50 per cent of the total emissions. Sulphur dioxide is a colourless gas with a sharp odour. Up to 70 per cent of its emissions come from combustion of emissions, fuel oil - about 15 per cent. Emissions containing impurities in the form of smoke, mist or vapour particles are called aerosols. The total number of varieties of aerosols polluting the atmosphere is in the hundreds. Aerosols have a detrimental effect on the ozone layer of the atmosphere [4].

To quantify the content of an impurity in the atmosphere, the concept of concentration is used - the amount of substance contained in a unit volume of air, reduced to normal conditions [5].

Atmospheric air quantity is a set of its properties determining the degree of impact of physical, chemical, biological factors on people, flora and fauna, as well as on materials, structures and the environment as a whole. Atmospheric air quality is considered satisfactory if the content of impurities in it does not exceed the maximum permissible concentration (MPC) - the maximum concentration of impurities in the atmosphere, attributed to a certain averaging time, which at periodic exposure or throughout a person's life does not have a direct or indirect impact on him and on the environment as a whole, including remote consequences. Direct impact is understood as temporary irritation of the human organism, causing odour, coughing, headache. When harmful substances accumulate in the organism above the specified dose, pathological changes of individual organs or the organism as a whole may occur. Under indirect impact are understood such changes in the environment, which, without having a harmful effect on living organisms, worsen the usual conditions of habitat: green areas are affected, the number of foggy days increases [6].

The main criterion for establishing MAC standards for assessing atmospheric air quality is the impact of airborne pollutants on the human body [7].

Two categories of MACs are established for the assessment of atmospheric air quality: maximum single dose (MACm.r) and average daily dose (MACc.s).

MACm.r is the main characteristic of hazard of a harmful substance. It is established to prevent reflex reactions in humans at short-term exposure to atmospheric impurities. This standard is used to assess substances that have an odour or affect individual sensory organs.

MACc.c - is established to prevent general toxic, carcinogenic, mutagenic and other effects of the substance on the human body. The substances assessed by this standard have the ability to temporarily or permanently accumulate in the human body [8].

By the beginning of 1999, about 1000 substances were assessed according to MPC norms, but dozens of new, poorly studied substances, most of which are harmful to humans, animals and plants, are added to this number every year. The list of substances, the content of which is rationed, therefore, is constantly replenished. Temporary standards of MPC of pollutants in the air for woody vegetation (MPCl) have been established [7].

If substances have a harmful effect on the environment in lower concentrations than on humans, then during rationing on the basis of the threshold of action of this substance on the environment. The impact of substances, for which MACs are not established, is assessed according to the approximate safe level of exposure to air pollutant (ESL) - a temporary hygienic standard for an air pollutant [8].

MPC norms for atmospheric air are single for the territory of a particular country; MPCs established in other countries may differ. For example, in the USA the MPC for SO2 is 0.75 mg/cm^3, and in Ukraine - 0.5 mg/cm^3. The established norms in each country are regulated by international health, environmental protection and various international organisations. For sanitary protection zones, resorts and recreation areas, MACs are set 20% lower than for residential areas [9].

Violations of the established norms are prosecuted by law, which provides for certain penalties. Such laws exist in every country, as it has been established that constant exceeding of the permissible concentration of at least one of the standardised substances leads to an increase in morbidity by 1.7 times, and in some age groups - up to three times. Atmospheric pollution also has a direct impact on structures and decorative ornaments, monuments, etc. In accordance with normative and technical documentation, environmental quality rationing is carried out in order to establish maximum permissible norms of environmental impact, which guarantees environmental safety and preservation of the genetic fund, ensures rational use and restoration of natural resources under the condition of sustainable development of economic activity [10].

For each projected and operating facility, which is a stationary source of air pollution, the standards of maximum permissible emissions (MPE) of pollutants into the atmospheric air are established. MPEs are established based on the condition that emissions of harmful substances from a given source in combination with other sources do not create a surface concentration exceeding the MPC outside the sanitary protection zone: C+Cf MPC, where

C - concentration of the substance in the surface layer from the design source, while maintaining the MPE norms;

Cf - background concentration of the same substance.

If, at a given enterprise or group of enterprises located in a given region, MPE values cannot be achieved immediately for objective reasons, a temporarily agreed emission (TCE) is established. The TPL standard is set for the period of development and organisation of air protection measures to ensure that the standards are achieved MPE. The validity period of MPE is set for 5 years. In case of new production facilities, reconstruction of existing ones, changes in the technological process or the type of raw materials used and other cases, the MPE standards are revised [11].

For each city, based on MPE standards of enterprises and background composition of atmospheric air, city-wide MPE standards are developed, according to which individual MPEs of enterprises can be revised downward [12].

Compliance with the established quality standards ensures a favourable environmental situation in the region in accordance with the requirements of the RK law on the environment. MPE is established for each stationary source based on the calculation that the aggregate emission from all sources of atmospheric air pollution, taking into account the development perspective, will not lead to exceeding the MAC standards in the surface layer [13]. MPE is established for conditions of full load of process and gas cleaning equipment and their normal operation. The MPE should not be exceeded in any 20 minute period of time. For small sources it is expedient to establish MPE from their aggregate with preliminary combination of them into an area or point source. The MPE is determined for each substance separately, including in case of summation of harmful effects of several substances [14].

Based on the results of calculation of MPE standards for each stationary source of emissions, the emission limit of enterprises as a whole is set, the MPE is set taking into account background concentrations of the energetically reliable maximum concentration [15]. It is a characteristic of atmospheric pollution and is defined as a concentration value, which is exceeded in no more than 6% of cases from the total number of observations. Background concentration characterises the total concentration generated by all sources located in a given area [16]. Establishment of MPE for a source is preceded by determination of its zone of influence. For enterprises and sources whose zones of influence are entirely located within the city, where the total concentration from all sources is less than the MPC. The emission values used in the calculations are taken as MPCs. A network of points and monitoring stations has been established to obtain information on the state of the air basin. Emissions inventory is regularly carried out - accounting of the main sources of atmospheric pollution, quantity and composition of emissions [17].

1.2 Anthropogenic environmental pollution and health status of the population

Environmental pollution caused by anthropogenic human activity has a certain impact on the formation of population health, especially in modern socio-economic conditions. In this regard, the problem of unfavourable influence of environmental factors on the state of human health becomes more and more relevant every year [18].

In the late 60s - early 70s, G.I. Sidorenko first set the environmental hygiene tasks related to the determination of quantitative dependencies in the system "environment - health". Later this science developed criteria and methods for quantitative assessment of the impact of environmental factors [19].

The contribution of anthropogenic factors to the formation of health deviations ranges from 10 to 60% [20].

Kazakhstan has a complex and unstable, and in some areas even acute unfavourable environmental situation. A rather serious unfavourable ecological and hygienic situation is also characteristic of some regions of Akmola oblast. The existing problem predetermines the fact that the issue of the impact of environmental pollution on public health has received a wide resonance in the literature [21].

Assessment of the significance of environmental pollution by biological responses of the human organism, by health indicators is more objective than comparison of concentrations of individual pollutants with hygienic standards, because it integrally takes into account the impact of all, including unidentified, pollutants, their complex and combined effect on the human body [22].

One of the leading factors of anthropogenic impact on health is aerogenic. In this case, the impact on the human body can be manifested mainly by three types of pathological effects.

1. Acute intoxication occurs when a toxic inhalation dose is administered at one

time. Toxic manifestations are characterised by an acute onset and pronounced specific symptoms of poisoning.

2. Chronic intoxication is caused by prolonged, often intermittent, ingestion of chemicals in subtoxic doses, begins with the appearance of low-specific symptoms.

3. Long-term effects of exposure to toxicants.

- Gonadotropic effect is manifested by influence on spermatogenesis in males and ovogenesis in females, as a result of which there are disorders of reproductive function of the biological object.

- Embryotropic effects are manifested by disturbances in the intrauterine development of the foetus:

- teratogenic effect - occurrence of organ and system disorders manifested in postnatal development;

- embryotoxic effect - death of the foetus, or reduction of its size and weight with normal tissue differentiation.

- Mutagenic effect - alteration of hereditary properties of an organism, due to DNA disorders.

- Oncogenic effect - development of benign and malignant neoplasms.

The results of medical-ecological and hygienic studies convincingly show that atmospheric air pollution causes certain manifestations of toxic reactions in the population, starting from the early stages of ontogenesis [23].

Aerogenic effects on the health status of the paediatric population.

The formation of child health disorders in the perinatal period is predominantly associated with the conditions arising in the mother during pregnancy and is caused by the influence of the maternal organism on the foetus and environmental pollution [24, 25]. It has been found that placentas of women living in conditions of increased atmospheric pollution have various signs of oppression of compensatory-adaptive mechanisms [18, 23]. Certain pollutants have the ability to penetrate the placental barrier [26, 27]. More than 600 chemical substances are known to be able to penetrate from mother to foetus through the placenta and to a greater or lesser extent negatively affect its development [21]. Therefore, disorders of embryonic development are closely related to this ability of xenobiotics, due to which the development of the embryo occurs under conditions of chemicalisation of its internal environment.

A statistically significant and consistent, as the level of air pollution increases, decrease in the weight and body length of newborns has been established [23]. An increase in the number of premature infants [28], the total share of low birth weight and large infants [29] has been found in polluted areas. Inhalation exposure to components of natural sulphur-containing gas condensate during pregnancy leads to the development of hypotrophy of fetuses of experimental animals [30]. After prenatal exposure to formaldehyde there is a decrease in the indicators of physical development of rats [31]. The influence of aerogenic pollution on anthropometric indicators at birth is also noted by other researchers [10, 11, 27, 32].

Long-term studies have revealed a correlation between birth weight indicators and levels of air pollution by various pollutants. A reliable direct correlation between the frequency of low birth weight births and concentrations of hydrogen sulfide and formaldehyde in the air at early stages of gestation and carbon monoxide at later stages of gestation has been observed [33].

A reliable direct correlation between the frequency of birth of large newborns and total exposure to sulphur and nitrogen dioxides at early stages of intrauterine development, as well as with exposure to benzapyrene and at later stages has also been established [34,35].Thus, atmospheric air pollutants have multidirectional effects.

As can be seen from Figure 1, the contribution of atmospheric air pollution to the formation of various anthropometric indicators of newborns ranges from 1.1% (head circumference) to 12.6% (body weight), and in the formation of disharmonious disorders of weight and height characteristics at birth reaches 16.8%. Air dustiness in residential areas is of the greatest importance [10].

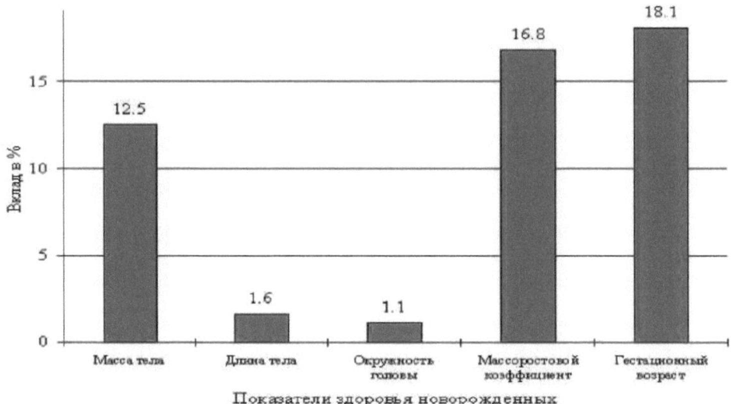

Figure 1. - Contribution of total atmospheric air pollution to the formation of newborn health indicators [15].

A number of studies have noted that the incidence of preterm labour is higher in environmentally unfavourable conditions [16, 36]. It has been found that pregnant women exposed to the combined action of chemical substances and physical factors have suppression of cellular and humoral immunity, as well as a high titre of antibodies against the tissues of the foetal egg and foetus, which indicates the depletion of "blocking" serum factors and accelerates the rejection reaction of the homograft [37].

A wide range of scientific studies have established that intense environmental pollution has an impact on the prevalence of congenital anomalies [10, 17, 28, 38].

Areas with higher levels of air pollution have been found to have a higher prevalence of multiple congenital malformations, congenital limb anomalies, and harelip [39].

Correlation analysis of atmospheric air pollution levels (dust, sulphur dioxide, nitrogen dioxide, sulphur oxide, carbon monoxide, hydrogen sulphide) and the prevalence of congenital anomalies in newborns revealed a reliable relationship only with nitrogen dioxide concentrations (r=0.72). At the same time, there was a significant direct correlation with the number of motor vehicles (r=0.98). It can be assumed that specific pollutants contained in its emissions play a role in this case [40].

The assessment of the negative impact of environmental pollution on the morbidity of the paediatric population is considered to be the most informative [39, 41].

Morbidity is the most characteristic, officially recorded response to harmful environmental effects, reflecting both long-term and chronic effects of a pollutant [42].

A number of studies have established a certain correlation between the level of morbidity in children under 1 year of age and the environmental situation, with the most frequently reported impact of atmospheric air pollution on the incidence of respiratory diseases [1, 26, 38].

There is no doubt about the relationship between chemical aerogenic exposure and various respiratory pathologies [19, 26, 43].

Allergic diseases are known to be among the leading ecopathological conditions [44].

Many industrial pollutants are inherently sensitising, and after adsorption on a protein carrier can acquire the properties of full-fledged allergens [39,45].

It has been established that as the technogenic impact on the environment increases, the specific weight of staphylococcal (resident) bacterial carriers among the population increases.

Aerogenic exposure affects the state and functioning of the cardiovascular system [39,46].

Oncopathology - as a criterion of unfavourable impact of anthropogenic atmospheric air pollution on the human body.

Many works are devoted to the study of carcinogenic effect of metals. Carcinogenic properties of nickel and hexavalent chromium compounds have been established by many researchers [43, 46, 47]. They are pulmonotropic carcinogens, as they most often cause lung cancer [48]. In addition, chromium pollution in the air basin can cause an increase in malignant neoplasms of the skin, lymphatic and haematopoietic systems [20]. Researchers have noted the role of cadmium in the formation of prostate tumours [48, 49, 50], also lung cancer [51]. The carcinogenicity of beryllium has been established [52].

The blastomogenic activity of fly coal ash has been determined, which may be due to the presence of components such as nickel, cobalt, beryllium, and chromium [53].

Carcinogenic properties have been identified in industrial oils [54].

It has been established that the incidence and mortality from lung cancer are higher in areas with a high content of benz[a]pyrene in the air [2, 29]. The association of oesophageal cancer with high levels of benz[a]pyrene and other polycyclic aromatic hydrocarbons in the environment has been established [55]. Proven carcinogens are polychlorinated biphenyls and dioxins [56].

Vinyl chloride causes angiosarcoma of the liver [57].

There is no doubt about the relationship between benzene content in the environment and the prevalence of leukaemia [39, 45]. The prevalence of this nosological form of cancer pathology is also associated with exposure to butadiene [58].

The link between stomach cancer and exposure to leaded petrol has been discussed [59]. The carcinogenic role of emissions from cars with diesel engines has been proven [60]. The carcinogenicity of acrolein has been discussed [52, 53]. There is limited evidence of a link between formaldehyde exposure and the occurrence of nasopharyngeal cancer and possibly nasal cancer [49].

One of the leading values in the formation of cancer morbidity in the urban population is radon [61, 62].

In addition to carcinogenic risk, there is the danger of toxic exposure to atmospheric air pollutants.

The high risk of toxic effects on the population living in the vicinity of major motorways is determined by the concentrations of acrolein and acetaldehyde [63].

1.3 Analysing the current state of the region's environment

1.3.1 Physical and geographical characteristics of the study area

Three districts of Akmola region (Akkol, Burabai, Zerendinsky) with different anthropotechnogenic load were chosen as objects of the study. Burabai and

Zerendinsky districts are residential areas, state national parks "Burabai" and "Kokshetau" are located here, the environmental situation in Akkolsky district is directly affected by chemical industry enterprises located in Stepnogorsk (Figure 2).

Figure 2. - Map of industrialisation of Akmola region [17]

Akmola region occupies an area of 146.219 thousand kilometres^2. Population - 748,326 thousand people as of 2014.

The region is rich in mineral resources, represented by significant reserves of the following minerals: gold, uranium-containing ores, iron ores, hard coal, construction materials [64]. Ores of deposits are usually complex in composition and contain, along with the main and associated components, harmful impurities, including environmentally dangerous, toxic elements or their mineral compounds: radioactive isotopes of uranium, thorium, potassium-40, arsenic, beryllium, selenium, phosphorus, antimony, asbestos and others [65]. There are more than 20 mining and processing enterprises in the Oblast [66].

The climate of Akmola region is sharply continental, characterised by hot, arid summers and severe winters. Continentality of the climate is manifested in large annual and daily amplitudes of air temperature fluctuations [67]. The average maximum temperature in July +190C, +210C, January -160C, -180C. A moderate amount of precipitation falls during the year, on average from 490 to 590 mm. Precipitation is characterised by low salinity (up to 10.0 mg/l). The pH value of precipitation on average for the year is - 6.0 with a norm of 6.0-9.0 [68]. There are isolated cases of acid precipitation (pH = 4.5). The average natural radiation background on the territory of the region is 0.25 μSv/hour with an increase in the northern part of the region in local areas up to 0.35 μSv/hour and with a general decrease in the southern part up to 0.15 μSv/hour [69]. On the territory of Akmola oblast radioactive waste is generated only at one enterprise - "Stepnogorsk Mining and Chemical Combine" LLP. Radioactive waste is generated as a result of enrichment of polymetallic ores and process solutions, mother liquors during heap leaching of ores and contaminated equipment.

The average wind speed was 4.3 m/s, the frequency of occurrence of surface air temperature inversions was 1-16%. The frequency of occurrence of stagnation of air masses was 13.8%, the frequency of occurrence of periodicity of inversions was 40%, the frequency of occurrence of fogs was 1.2%.

All this allows us to characterise the climatic regime of the area as favourable for self-purification of the atmosphere [70,71].

1.3.2 Atmospheric air condition

The main sources of pollutant emissions in Akmola Oblast are heat and power enterprises and motor transport. The largest stationary sources of atmospheric pollution are TPP "Jet-7" LLP in Stepnogorsk city and State Enterprise "District Boiler House No. 2" in Kokshetau city. Motor transport emits 42.1% of the total amount of pollutants into the environment [64].

Total air pollutant emissions from stationary and mobile sources in 2014 amounted to 224.75 thousand tonnes, which is 35.67 (15.9%) more than last year (Table 1). This increase is based mainly on the growth of industrial production in the region, intensified activity of enterprises, increasing their production capacity, as well as the completion of the campaign to involve in the accounting of emissions from the private residential sector in Akmola region. In addition, in 2014 accounting covered 5,250 units of stationary sources, while in 2013 - 2,423 units. This is a consequence not only of better accounting work, but also due to the emergence of additional stationary sources at enterprises [65].

Table 1. - Data for 2013 and 2014 (tonnes)

Indicators	2013	2014
Total emissions	189072,43	224748,47
Emissions from stationary sources	96878,45	130087,78
incl.: solid	58468,57	60277,23
gaseous	38409,88	69810,55
Emissions from mobile sources	92193,98	94660,69

The mass of air pollutant emissions from motor transport changed insignificantly, by 2.7% more than last year. At the same time, the total number of motor vehicles registered by the Road Police Department of Akmola region totalled 98305 units in 2013 and 103746 units in 2014.

Emissions from stationary sources increased by 34.3%. The reasons for the growth of emissions from stationary sources are outlined above.

Emissions of solid substances have relatively stabilised. Their growth for the whole of 2013 was only 3%. This is due to the fact that nature users, forced to do so by the terms of environmental impact assessments and permits for special nature use, as well as by the instructions of environmental protection inspectors, install ash collection equipment at their emission sources.

In 2014, the number of enterprises with relatively large volumes of environmental pollution increased, which makes it possible to classify them into the group of enterprises that must obtain environmental permit directly from the MEP. Such enterprises include: LLP "Orken-Atansor"; branch of JSC "SSGOPO" Alekseevsky dolomite mine; LLP "Nefrit-SV" [66].

1.3.3   Status of land resources and production and consumption waste

A significant share of solid waste on the territory of the region is industrial waste, which is mainly represented by mining wastes in summer and ash and slag wastes in winter [65].

Comparison of data on waste generation volumes for 2014 (8846.19 thousand tonnes) and 2013 (2385.04 thousand tonnes) shows an increase of 3.7 times (Table 2).

Table 2. - Comparative data on waste generation for 2013 and 2014 (in thousand tonnes)

Indicators	2013	2014
Total waste	2385,04	8846,19
Including:		
3 classes + litter	192,80	140,45
stripping	669,51	6984,76
ash and slag	614,85	816,05
MSW	907,88	904,93

The main reason for the increase is the relative growth of industrial production in Akmola region, and in particular - the intensification of mining and mining processing enterprises, accompanied by a real increase in the volume of overburden formation. The completeness of the reflection of the formed overburden in the reports of enterprises has increased, as well as the accuracy of the accounting itself [72]. If in 2013, 669.51 thousand tonnes of overburden and off-balance sheet ores were formed and reflected in the reports of enterprises, in 2014 - 6984.76 thousand tonnes, i.e. 10.4 times more. Overburden rocks at the enterprises of Akmola region are stored in rock dumps, used to fill roads in the territories of enterprises, as well as for the formation of heap leaching plants for gold [67].

Ash and slag in 2013 was formed 614.85 thousand tonnes, and in 2014 - 816.05 thousand tonnes, i.e. by 32.7% more. This growth is explained, as it was already indicated, by the activation of industrial enterprises, in particular, the largest owners of boiler houses - "Jet-7" LLP, "SGHK" LLP and SCP at PCW "RK-2" [65]. [65].

Hazard class 3 wastes in Akmola region are mainly represented by tailings of gold processing plants of MMC Kazakhaltyn OJSC. In 2013, these wastes were generated in the amount of 177.56 thousand tonnes, and another 3.27 thousand tonnes of cooling lubricants of Stepnogorsk bearing plant and oil sludge of "Altyntau

Kokshetau". Total Class 3 waste in 2013 was generated 180.83 thousand tonnes [73]. Whereas in 2014 the following was generated: 120.65 thousand tonnes of gold mining tailings (OJSC "MMC Kazakhaltyn" for the first 9 months of the year worked only at one third of its production capacity), 6.06 thousand tonnes of coolants and oil sludge, total class 3 waste - 126.71 thousand tonnes, i.e. 30% less than last year. This decrease is primarily due to the fact that only one mine out of three mines of MMC Kazakhaltyn OJSC operated during the first three quarters of 2014. Tailings ponds exist on the balance sheet of the enterprises to store Class 3 wastes [68].

Poultry manure generated at Akmola Phoenix JSC's poultry farm, which is classified as hazard class 5, was 11.97 thousand tonnes in 2013 and 13.74 thousand tonnes in 2014, i.e. 14.8% more.

In 2013, 907.88 thousand tonnes of municipal solid waste (MSW) was generated and disposed to landfills, and in 2014 - 904.93 thousand tonnes, i.e. almost the same amount. MSW is not utilised on the territory of Akmola region due to the absence of a waste processing plant. MSW and a part of ash and slag are transported to landfills, the number of which in the region exceeds 600. Ash and slag from large enterprises (LLP "Jet-7", LLP "SGHK", SCP at PCW "RK-2") are stored in special ash dumps [69].

Out of 648 waste accumulators 118 are on the balance of enterprises, 530 - on the balance of local administrations [65].

There are 3979 natural users polluting the environment (including peasant farms with only motor transport as emission sources) in Akmola region [70].

According to the data of the Akmola Department of Environmental Protection, there are 14458.5 ha of disturbed lands in the region, of which only 99.97 ha have been rehabilitated [74].

The main problems affecting the state of land resources are:

- insufficient reclamation of disturbed lands;

- rural landfills are not used to bury rubbish, which creates conditions for its spreading around the neighbourhood and the formation of unauthorised dumps;

- There are no production and consumption waste recycling facilities.

The main sources of land pollution in many districts of the Oblast are unauthorised dumps and abandoned quarries.

In many districts there is no reclamation of these disturbed lands, for 2014 in Zhaksyn, Atbasar, Enbekshilder and Tselinograd districts 24 ha of land disturbed as a result of anthropogenic activity were reclaimed [75].

According to the data of the Akmola Regional Department of the Agency of the RK on Land Resources Management as of 01.10.2014 (working statement to form No. 22) on the lands of settlements disturbed lands are: in Zerendinsky district - 653 ha, Ereimentau - 421 ha, Kokshetau city - 417 ha, Sandyktau - 370 ha, Bulanda - 248 ha, Atbasar - 247 ha. In total, there are 3238 ha of disturbed lands on the lands of settlements in the region [71].

The disturbed lands of the PAs occupy 15 hectares and all of them are located in Burabai district.

The main factors affecting the state of land resources are:

- No reclamation of disturbed lands is carried out;

- The soil and fertile layer is not completely removed during reconstruction and construction of motorways during the construction of bypass roads, clay quarries;

- Rural landfills are not being buried, resulting in an annual increase in the size of the landfills;

- Formation of unauthorised landfills.

- Lack of production and consumption waste recycling facilities.

1.3.4 Condition of subsoil, surface and ground waters

Akmola region has a number of large deposits not only in the scale of Kazakhstan, but also in the whole world, such as: Vasilkovskoye gold deposit, technical diamond deposit "Kumdykol" - the only one in Kazakhstan, Kuletskoye deposit of fine-checkered muscovite and uranium deposits. The region also has significant reserves of commonly occurring minerals [68].

Mining has a diverse anthropogenic impact on the natural environment. The changes occurring in it often negatively affect the state of ecological systems [65].

Alienation and disturbance of land over significant areas, extraction of rocks, minerals, underground water from the subsoil, disposal of solid and liquid wastes from processing and concentration of minerals, creation of accompanying infrastructure (additional industries and residential settlements with their harmful waste water and domestic waste) lead to changes in the condition and properties of rock massifs and underground hydrosphere, affect landscapes, surface water, soil, atmosphere and through them (as well as directly and directly).

Ores of deposits along with the main and associated components contain harmful impurities: arsenic, beryllium, selenium, phosphorus, antimony, asbestos, sulphur and others [76].

The main task of reducing the negative impact of the consequences of the development of mineral deposits on the environment is to ensure control over compliance by subsoil users and state bodies with the legislation of the Republic of Kazakhstan.

On the territory of the Oblast there are 142 objects of the State Control on protection of subsoil of commonly occurring minerals and about 20 mining and processing enterprises [72].

Within Akmola Oblast, the largest rivers are Ishim, Koluton, Zhabai, Seleti, Nura, Shaglinka, Kylshakty, and Ters-Akkan. The rest of the rivers have a small length, most of them dry up in summer [73].

By their regime, the rivers belong to the type of plain rivers, predominantly snow-fed. The rivers open in mid-April. The water in floods is turbid, odourless, with low oxidability. Due to dilution by melt water, the content of calcium and magnesium salts decreases and hardness decreases. The highest mineralisation rates of total hardness are observed in June. In some severe winters, some small rivers freeze to the bottom and water flow temporarily stops [74].

The main water artery is the Ishim River. The population of rural areas uses the water of the river for household and drinking purposes, both centralised and decentralised. The Astana (Vyacheslavskoye) reservoir is located in the upper reaches of the river. The catchment area of the Ishim River in the region is 84.3 km^2, the annual runoff volume at 90% availability is 129.97 million m^3/year [75].

Also a large river of Akmola region is the Nura river. Nura. Its length is 406 km and its catchment area (within Akmola oblast) is 9460 km^2. The annual runoff volume at 90% availability at the mouth of the river is 66.4 million m^3/year.

The Shaglinka River flows in the northern part of the oblast, which is the main water source of the oblast centre - Kokshetau city. The length of the river on the territory of the oblast is 144 km, the average annual flow volume is 40.77 million m^3/year.

There are about 140 large lakes and a large number of small lakes with a water surface area of less than 1 km^2 on the territory of the region. The largest lake is Lake Tengiz. The area of water mirror is 1590 km^2. Another large lake is Kurgalzhino, with a water surface area of 330 km^2 [77].

The depths of lakes are usually shallow. Their average depth does not exceed 1.0 - 1.5 metres. Due to the flat relief of the bottom of lake basins the size of lakes in summer time sharply decreases. The majority of small lakes dry up, some of them retain only separate channels. A significant number of lakes in the region is in the stage of degradation due to siltation and gradual overgrowing of their hollows with aquatic vegetation [71].

However, the most difficult situation has developed in the Shchuchinsko-Borovsky resort zone (hereinafter referred to as the Shchukchye resort zone), where, as a result of irretrievable water withdrawal, there is an annual decrease in the volume of water in lakes (the most typical example is Lake Shchuchye).

Due to the negative water balance, the ShchBKZ lakes have been shallowing. Since 1986 the water level has been annually decreasing by 15 - 20 cm. Thus, the level of Lake Shchuchye decreased by 2.18 m, Lake Bolshoye Chebachye by 1.5 m, Lake Maloye Chebachye by 0.7 m. At the same time, the condition of surface water bodies of the ShchBKZ is not monitored to a sufficient extent. At present, according to RSE "Kazgidromet", gauging stations operate on lakes Borovoye, Bolshoye Chebachye and Shchuchye [75].

One of the problems of Shchukhinsk is water supply. For household drinking needs of Shchuchinsk city water in the amount of 1.3 million m^3 per year is taken from lake Shchuchye. This leads to a drop in the water level in the lake. In the resort area, 17 underground wells are used for water supply. For water supply of the Shchuch'e resort zone in 1968 - 1970 the search of underground water deposits was carried out. Hydrogeologists of the Sinegorsk hydrogeological expedition discovered groundwater deposits with a total operational reserve of 411 litres per second. Practically all these reserves are currently not exploited, although the exploitation of the Shchuchinsk and Verkhne-Kylshaktynskoye groundwater deposits with the involvement of groundwater reserves of the Kazgorodskoye deposit could completely solve the problem of water supply to the town of Shchuchinsk with the exclusion of water intake from Lake Shchuchye [76].

Factors affecting the qualitative and quantitative composition of groundwater and surface water are: irretrievable water withdrawal, pollution and clogging of water bodies, area washout of the top layer from arable land located in the catchment area. The development of blue-green algae leads to intensive overgrowing and other aquatic vegetation of water bodies. The main polluters of lakes are settlements and industrial enterprises located in their catchment areas [77].

Poor technical condition of water supply systems, pollution of surface water sources create one of the most urgent problems - providing the population with quality drinking water [78, 79].

Water supply to the population and production facilities of the oblast has been negatively affected by the fact that in recent years the Sergeevsky, Seletinsky, Yaroslavsky, Yablonevsky, Kiyminsky, Derzhavinsky, Zerendinsky group water pipelines have ceased to function.

Intensive pollution of water sources, poor sanitary and technical condition of water supply facilities and networks, high microbial contamination of drinking water in district centres and rural settlements are the causes of epidemiological complications and high incidence of acute intestinal infections [80].

In 2014, compared to 2013, there was a decrease in water withdrawal. Thus, in 2014, the total water withdrawal was 127.638 million m^3, which is 3.497 million m^3

less compared to the same period of 2013 (Table [3].

Table 3. - Water withdrawal for drinking and technical needs (million cubic metres)

Name of need	2013	2014
Water withdrawal, total	131,135	127,638
Including: superficial	62,493	67,759
underground	42,671	40,919
Transport losses	25,971	18,960

The decrease in water withdrawal was primarily due to a reduction in transport losses.

For 2014, there was a slight decrease in groundwater use, while surface water abstraction increased.

Despite the implementation of measures to improve water supply to settlements (replacement of main and branch pipelines), water losses during transportation are still high. Total water losses in 2014 amounted to 18.96 million metres[3], as many main *and* branching water pipelines have fallen into disrepair due to a long period of operation.

Basically, on the territory of the region, pollution of surface and underground water sources occurs due to the discharge of insufficiently treated and untreated wastewater into storage tanks and filtration fields. Out of 38 treatment facilities with a total design capacity of 83.6 million m^3/year, only 7 facilities are operating [81].

Wastewater discharge volumes for 2013 and 2014 are presented in Table 4.

Table 4. - Wastewater discharge volumes (mln. m^3)

Indicators	2013	2014
Total, industrial water disposal	22,032	37,49
Including into surface water bodies	6,09	10,669

The increase in the volume of wastewater discharge is associated with the increase in the production of products requiring the use of both technical and drinking water, which, having become contaminated in the process of use, passes into the category of wastewater. An important factor influencing the increase in wastewater discharge is the growth of population living in cities and other settlements.

The following water outlets for wastewater discharge into surface water sources in Akmola Oblast are available: treatment facilities in Stepnogorsk (Jet-7 LLP), and storm water drainage facilities at the industrial site of the former Progress JSC (Krodako LLP). Wastewater from the above-mentioned facilities is discharged into the Aksu River after being treated at treatment facilities. The results of observations show that below the discharge of wastewater from these enterprises in the river increases the content of salt ammonium, nitrates, suspended solids, but they do not exceed permissible standards. The qualitative composition of water in the river downstream of discharges fully meets the requirements for water bodies of sanitary and domestic use [79].

Water discharge of Kokshetau city is carried out to biological treatment facilities with a design capacity of 32 thousand m^3/day. Only 18 thousand m^3/day is subjected to biological treatment [82].

The issue of wastewater treatment is irrationally solved, as wastewater passing through the treatment plant is mixed with untreated wastewater and discharged into the Myrzakolsor reservoir, which was accepted as a temporary option for wastewater accumulation with a design volume of 49 million metres^3. As a result of the construction of a barrier dam, the volume of the reservoir was increased to 120 million metres^3. The actual volume accumulated was 110 million m^3. In order to avoid overfilling of the Myrzakolsor reservoir, KNS-13 transfers wastewater to the Akhmetzhansor reservoir, which has a design capacity of 20 million m^3 and actual accumulation of 15 million $m^{(3)}$) [83].

The issue of constructing bioponds to provide deep additional treatment of wastewater in order to solve the issue of its further utilisation is not addressed. This would solve the problem of wastewater utilisation and would not pose a threat of emergency situation at storage tanks [84].

Wastewater treatment in Shchuchinsk is carried out at mechanical and full biological treatment facilities with a capacity of 16.5 thousand m^3 per day. Treated wastewater is discharged into the "Balykty" storage tank with a capacity of 14.5 million m^3.

Wastewater from the Aidabul distillery after treatment is discharged into a wastewater storage tank.

Mechanical wastewater treatment facilities of "Shchuchinskaya poultry farm" LLP, the design capacity of which was 2.5 thousand m^3/day, completely fell into disrepair and were written off the balance sheet of the enterprise.

Treatment facilities of mechanical and biological purification of the Republican Youth Camp "Okzhetpes" are in an emergency condition and currently do not function. Wastewater is transported by special vehicles to the storage pond without treatment [75].

The sewage treatment plants located at railway stations are mothballed due to small volumes of effluent insufficient for normal operation of the treatment facilities.

Wastewater from district centres is also discharged without treatment into storage tanks.

The lack of centralised sewerage systems entails the construction of numerous local wastewater collectors (cesspools), which often creates a threat of overfilling and environmental pollution [85].

CHAPTER 2

METHODS OF COMPREHENSIVE ASSESSMENT OF THE DEGREE OF
ENVIRONMENTAL STRESS IN THE STUDIED REGIONS

To assess the quality of the environment in the studied districts of Akmola region, studies were conducted and available data on air, water and soil pollution were analysed.

2.1   Determination of the main indicators of atmospheric air pollution

The content of phenol, sulphur dioxide, carbon monoxide, nitrogen dioxide and formaldehyde was determined in the atmospheric air.

The main quality criteria are the values of maximum permissible concentrations (MPC) of pollutants in the air of residential areas.

Determination of daily average concentration of phenol in atmospheric air.

For atmospheric air, the average daily MAC of phenol is 0.003 mg/cubic metre; the maximum single MAC is 0.01 mg/cubic metre, hazard class 2 [86].

Measurement method. The method is based on absorption of phenol from air by sodium carbonate solution, extraction of impurities with hexane, extraction of phenol from the resulting solution with butyl acetate, its re-extraction into aqueous sodium hydroxide solution and measurement of mass concentration after acidification of the re-extract. During the measurement process, phenol fluorescence is excited, recorded and the phenol content is automatically calculated using a calibration characteristic stored in the analyser memory [87].

Air sampling. When analysing the atmospheric air of residential areas, sampling is carried out in accordance with GOST 17.2.3.01. The sampling time is 25 - 30 min. by aspiration into two consecutively connected absorption vessels filled with 5 cubic cm of absorption solution; volume flow rate is 2.0 cubic dm /min. A total of 50 cubic dm of air is sampled.

Storage period of absorption solutions in closed vessels after sampling is 12 h.

In a separating funnel with a capacity of 50 cubic centimetres place 10 cubic centimetres of butyl acetate and add 10 cubic centimetres of sodium hydroxide solution according to 5.1.3. After thorough mixing and separation of layers the upper one is discarded and the lower aqueous layer is poured into a dry beaker with a capacity of 25 - 50 cubic centimetres, add hydrochloric acid solution according to 5.1.2, controlling pH with a universal indicator. The required pH value is 3 - 6. The obtained solution (control solution) is analysed on the device "Fluorat-02" [88]. [88].

Preliminary calibration of the instrument is carried out.

For analysers of modifications "Fluorat-02-1" and "Fluorat-02-3" the setting of the "Background" mode is carried out using distilled water, and the "A" parameter in the "Graduation" mode - using phenol solution with concentration of 1 mg/dm (item 5.1.5), having previously brought the pH of this solution to 3 - 6 with hydrochloric acid solution according to item 5.1.2. The "C" parameter is set equal to 1,000.

Determine the phenol concentration in the control solution in the "Measurement" mode. If the measured value exceeds 0.02 mg/cubic dm, the solvent must be purified.

For this purpose in a separating funnel with a capacity of 1000 cc shake 700 - 750 cc of solvent with 50 cc of sodium hydroxide solution according to 5.1.3 for 3 min. The pH of the bottom layer is checked using universal indicator paper. If the reaction of the medium is strongly alkaline (pH > 10), the solvent is washed with distilled water in portions of 50 cc until the wash water reaches a neutral reaction.

The solvent is then dried over anhydrous calcium chloride and distilled, collecting a fraction boiling at 124 - 126 °C [86].

Performing measurements

The absorbent solution from the Richter absorbent vessel is transferred to separating funnel with a capacity of 50 cc, the vessel is washed with distilled water and the wash water is added to the basic solution. The solutions belonging to one sample (in case of sampling in two consecutively connected vessels) are combined.

Add 10 cc of hexane and extract for 1 min. After separation of the layers, the lower aqueous layer is transferred to another separating funnel. The hexane layer is discarded. To the solution in the separating funnel add dropwise hydrochloric acid solution according to 5.1.2 to pH 3-6 (caution: foaming is possible!), add 10 cc of butyl acetate and extract for 1 min. The aqueous layer is discarded and 5 cc of sodium hydroxide solution according to 5.1.3 is added to the organic layer by pipette and re-extracted for 1 min [86, 87].

The obtained re-extract is placed in a dry beaker, check the pH of the solution by the universal indicator (should be > 10) and acidify with hydrochloric acid according to 5.1.2. Check the pH of the solution by the universal indicator (should be 3 - 6) and transfer it to a cuvette. Measure the phenol content in the sample in the "Measurement" mode.

Processing of measurement results

Mass concentration of phenol in the sample (X, mg/cubic metre) is calculated according to formula (1):

$$X = \frac{Q}{V_o},\qquad(1)$$

Where:

Q - phenol content in the absorption vessel, µg;

V_0 - volume of air sampled for analysis and reduced to normal (atmospheric air survey; pressure 760 mmHg,

temperature O °C) or standard conditions (working area air survey; pressure 760 mmHg, 20 °C), cu. dm.

In turn, Vo is found by formula (2):

$$V_0 = G \times \frac{P}{273 + t} \times U \times \text{Тау}\qquad(2)$$

Where:

t - air temperature at sampling (at the aspirator inlet), °C;

P - atmospheric pressure at sampling, mmHg or kPa;

G is a conversion factor equal to 0.359 when measuring pressure in mmHg. When measuring pressure in kPa, it is equal to 2.9;

U - air flow rate at sampling, cubic dm/min;

tau - sampling duration, min.

If the sampling device directly records the air volume (V, cubic dm), the product U tau is replaced by V in the above formula [88].

Determination of daily average concentration of sulphur dioxide and carbon monoxide in atmospheric air.

To determine sulphur dioxide and carbon monoxide in atmospheric air, the ultraviolet fluorescence method is used using an automatic gas analyser. The output signal for the analysed air sample is recorded.

Determine the appropriate concentration (using the calibration function).

The measurement result shall be expressed in micrograms per cubic metre or in milligrams per cubic metre or in equivalent volume fractions:

$d \pm D$ at confidence level P=0.95,

where $\pm D$ is the value of the error characteristic of the measurement result, specified in the documentation for a particular type of automatic gas analyser or given in the normative document establishing the methodology for measuring the mass concentration of sulphur dioxide using this type of automatic gas analyser.

The following formula (3) is used to convert millionths (ml/m$^{3)}$) to milligrams per cubic metre (mg/m$^{3)}$)

$$\rho_1 = \frac{\varphi_2 64 \cdot 273p}{22{,}407T \cdot 1013},\qquad (3)$$

where g1 - mass concentration of SO2 (CO), mg/m^3;

j$_2$ - volume fraction of SO2 (CO), ml^{-1} (ml/m$^{3)}$);

64 - molar mass of SO2 (CO), g/mol;

273 - standard absolute temperature, K;

p - measured gas pressure, hPa;

22.407 is the volume of 1 mole of gas at temperature 273 K and pressure 1013 hPa, l;

T - measured temperature, K;

1013 - standard gas pressure, hPa [86-88].

Determination of daily average concentration of nitrogen dioxide in atmospheric air.

The method for determination of nitrogen dioxide content in air with the Griss-Ilosvay reagent is based on the interaction of nitrogen dioxide and sulfanilic acid to form a diazo compound, which reacts with a-naphthylamine to produce an azo dye. The latter colours the solution from pale pink to red-violet. By the intensity of colouring of the solution the amount of NO_2 is determined. The sensitivity of determination is 0.1 µg in the analysed sample volume.

To determine the single NO_2 concentration, the tested air is drawn through a Richter absorber filled with 6 ml of absorbing solution at a rate of 0.25 l/min for 20 min. In the laboratory, the level of solution in the absorbent apparatus is brought to the 6 mL mark with distilled water. For analysis, 5 ml of the solution from each sample is transferred to a test tube and 0.5 ml of the compound reagent is added. The contents of the test tubes are shaken thoroughly and after 20 min (just before measurement) 5 drops of 0.06% Na_2SO_3 solution are added to the test tubes and shaken again.

The optical density is measured in 10 mm cuvettes at a wavelength of 540 nm relative to water. The time from the addition of the compound reagent to the measurement of the optical density of all samples must be the same.

The amount of NO_2 in the samples is determined from the calibration graph. At the same time the optical density of the zero sample is measured.

Calculation of nitrogen dioxide concentrations in the air is carried out according to formula (4):

$$C = \frac{a \cdot m}{V_0 \cdot b}, \qquad (4)$$

Where:

a - total sample volume in the absorption apparatus (6 ml);

b is the volume of the sample to be analysed (5 ml);

m - amount of NO2 in the sample, found from the calibration graph, µg; V_0 - volume of drawn air, reduced to normal conditions, litres.

Determination of daily average concentration of formaldehyde in atmospheric air.

Sampling of atmospheric air of populated areas is carried out by two consecutively connected absorption vessels, in which 5 cc of diluted absorption solution is placed. The sampling time is 20 min. at aspiration with a volume flow rate of 0.2 - 0.25 cubic dm/min.

Storage period of absorbent solutions in closed vessels is not more than 24 h. A series of analyses is carried out with the same absorption solution. The storage period of collected samples is not more than one day [89].

Measurements. The solutions from each absorption vessel are transferred into a test tube with a lapped stopper with a capacity of 10 cc. The tube is placed in a water bath and heated for 45 min. at 60 °C (sample solutions should be heated simultaneously with solutions for calibration of the analyser).

The solutions are cooled and the formaldehyde content of each absorption vessel is measured in the "Measurement" mode.

Processing of measurement results. Mass concentration of formaldehyde in the sample (X, mg/cubic metre) at sampling in absorption vessels is calculated by formula (5):

$$X = \frac{q_1 + q_2}{V_0}, \qquad (5)$$

Where:

q - formaldehyde content in the first absorption solution, µg;

q - formaldehyde content in the second absorption solution, µg;

Vo - volume of air sampled for analysis and adjusted to normal (atmospheric air study; pressure 760 mm Hg, temperature 0°C) or standard conditions (working area air study; pressure 760 mm Hg, 20°C), cubic dm.

In turn, Vo is found by the formula:

$$V_0 = G \times \frac{P}{273 + t} \times U \times \text{Tay} \qquad (6)$$

Where:

t - air temperature at sampling (at the aspirator inlet), °C;

P - atmospheric pressure at sampling, mmHg or kPa;

G is a conversion factor equal to 0.359 when measuring pressure in mmHg. When measuring pressure in kPa, it is equal to 2.9;

U - air flow rate at sampling, cubic dm/min;

tau - sampling duration, min [86-89].

2.2   Determination of the main indicators of surface water pollution

Determination of organoleptic indicators of water quality.

Odour Determination. The nature and intensity of an odour is determined by the sensation of the perceived odour. There are two groups of odours: naturally occurring odours and artificial odours. Typically, the nature of water odour is described by the terms presented in Table 5.

The odour intensity of drinking water is assessed using the 5-point system presented in Table 6.

Course of Determination. The odour of water is determined at room temperature and at 60 °C.

Determination of odour at room temperature. A flask with a lapped stopper with a capacity of 200-300 cm³ is filled to 2/3 of the volume of the water under test and shaken vigorously, then open the stopper and inhale its odour.

Table 5 - Odour character assessment scale

Symbol	Nature of the odour	Approximate genus of odour
A	Aromatic	Cucumber, floral
Б	Swampy	Silted, tinny.

Г	putrefactive	Fecal, sewage
Д	Wood	Wet wood chips, wood bark.
З	earthy	Stale, freshly ploughed earth, clay.
П	Mouldy	stale
p	Fishy	Fish oil, fish oil
C	hydrogen sulphide	Rotten eggs
T	Grassy	Grass clippings, hay
H	Undefined	Naturally occurring, not fitting the previous definitions

Kalitsun V.I., Laskov Y.M. Laboratory practice on water disposal and wastewater treatment WOD [86]

Table 6 - Assessment of odour intensity

Odour intensity	The nature of the odour manifestation	Odour intensity, points
No	The odour is not perceptible	0
Very weak	The odour is not perceptible to consumers, but is detectable in a laboratory test	1
Weak	The odour is noticed by the consumer if you draw their attention to it	2
Notable	The odour is easily noticed and causes disapproval of the water	3
distinct	The odour draws attention to itself and makes you refrain from drinking	4
Very strong	The odour is so strong that it makes the water unfit for consumption	5

Kalitsun V.I., Laskov Y.M. Laboratory workshop on water disposal and wastewater treatment [86]

Odour detection at 60 °C. To increase the intensity of odours, water is heated. A conical flask with a capacity of 200-300 ml is filled to 1/2 of its volume with water, covered with an hour glass and heated to 60 °C. Then the flask is shaken with a rotary movement and, by moving the glass, quickly determine the nature and intensity of the odour.

Definition of flavour. Drinking water should have a pleasant, refreshing taste without any off-flavour.

The taste of water depends on the mineral composition of the water, its temperature and dissolved gases.

There are four basic taste sensations: salty, sour, sweet, bitter. All other taste sensations are called flavours (alkaline, metallic, chlorine, astringent, etc.).

Determination of taste and flavour is carried out in known safe water at a temperature of 20°C. In doubtful cases the water is boiled for 5 min and cooled.

Course of the determination. Tested water is taken into the mouth in small portions, without swallowing, and hold 3-5 seconds. The presence of taste (salty, bitter, sour, sweet) or aftertaste (alkaline, glandular, metallic, astringent, etc.) is noted and their intensity is determined according to a 5-point system similar to that for odour (Table 6): 0 points - no flavour; 1 point - very weak; 2 points - weak; 3 points - noticeable; 4 points - distinct; 5 points - very strong.

Determination of water colour. Colour is a natural property of water due to the presence of humic substances, which are formed during the destruction of organic compounds and give it a yellowish to brown colour.

Water colour is determined photometrically by comparing samples of test water with standards that simulate the colour of natural water.

Course of Determination.

1. Preparation of the colour scale

To prepare the colour scale mix solutions No. 1 and No. 2 in Nessler cylinders in the following ratios specified in Table 7. The scale is stored in a dark place 1-2 months.

After measuring the optical density of the solutions, a graph is plotted with degrees of chromaticity on the abscissa axis and the optical density of the solutions on the ordinate axis.

2. 100 cm^3 (or 10 cm$^{3)}$ of filtered test water is measured into a Nessler cylinder, viewed from above against a white background and compared with the colour scale.

Table 7 - Ratio of solutions for preparation of colour scale (chromium-cobalt colour scale)

Solution No. 1, ml	0	1	2	3	4	5	6	8	10	12	14
Solution No. 2, ml	100	99	98	97	96	95	94	92	90	88	86
Degrees colour	0	5	10	15	20	25	30	40	50	60	70

Kalitsun V.I., Laskov Y.M. Laboratory workshop on water disposal and wastewater treatment [86]

3. When determining chromaticity using FEC, the optical density of a filtered water sample is measured in a blue light filter at 413 nm in a cuvette with a light absorbing layer thickness of 5-10 cm. Distilled water serves as a control. From the obtained optical density on the calibration graph the value of chromaticity in degrees is found.

The colour of drinking water should not be higher than 20°, in which case it is considered practically colourless. In individual cases, the State Epidemiological Service may permit the use of water with a colour of up to 35° [86].

Determination of the hydrogen index (pH) of water.

The hydrogen value of water (pH) - characterises the active reaction, determines the natural properties of water and is an indicator of pollution. Natural water usually has a slightly alkaline reaction. An increase in alkalinity indicates pollution or blooming, while an acidic reaction indicates the presence of humic substances (swamp water) or industrial wastewater. The hydrogen value of drinking water (pH) should be within the range of 6.09.0.

Course of Determination. Qualitative reaction (pH) is determined by the indicator. A strip of indicator paper is immersed in a flask with the tested water and its colouring is compared with the scale standards of the universal pH indicator [88].

Determination of nitrogen-containing substances.

The indicator of water pollution by organic substances of animal origin are salts of ammonia, nitric and nitric acids. The content of ammonium salts above 0.1 mg/dm^3

indicates fresh water pollution, because ammonia is the initial product of decomposition of organic nitrogen-containing substances.

Nitric acid salts (nitrites) are products of ammonia oxidation under the influence of microorganisms in the process of nitrification. Presence of nitrites in amounts exceeding 0.002 mg/dm^3 indicates the age of pollution.

Nitric acid salts (nitrates) are the end products of mineralisation of organic matter. The presence of nitrates in water without ammonia and nitrites indicates the completion of the mineralisation process.

Simultaneous content of ammonia, nitrites and nitrates in water indicates a clear unfavourable water source, permanent pollution. Elevated amounts of nitrite and nitrate without the presence of ammonia indicate the cessation of pollution at present. The presence of ammonia and nitrite in the water indicates the recent appearance of a permanent source of pollution. The presence of nitrates alone in the water indicates the end of mineralisation processes.

The permissible nitrate content in drinking water is not more than 45 mg/dm^3.

The chemical investigation starts with qualitative reactions, so as not to waste time quantifying those salts that are not present in the water.

Determination of nitrogen of ammonium salts. The method is based on the ability of ammonia and ammonium ions to form a compound with Nessler's reagent (Merkur ammonium iodide), which colours the solution yellow-brown. The intensity of the colouring is proportional to the ammonia content of the water.

Qualitative determination of ammonium salts with approximate quantification.

Determination procedure. In a test tube of colourless glass with a flat bottom diameter of 13-14 mm pour 10 cm$^{(3)}$ of water under study, add 0.2 cm^3 of segnet salt (potassium sodium tartaric acid) and 0.2 cm$^{(3)}$ of Nessler's reagent. After 10 min, an approximate determination of ammonium nitrogen is made according to Table 8.

Table 8 - Approximate determination of ammonium nitrogen

| Colouring on review | | Ammonia nitrogen content, mg/dm^3 |
On the side	From top to bottom	
No	No	Less than 0.04

No	Extremely faint yellowish	0,08
Extremely faint yellowish	Faint yellowish	0,2
Very faint yellowish	Yellowish	0,4
Light yellowish	Yellow	2,0
Yellow	Intensely yellow-brownish	4,0
Cloudy, sharply yellow	Brown. The solution is cloudy	8,0
Intensely brown. The solution is turbid	Brown. The solution is cloudy	20,0

Kalitsun V.I., Laskov Y.M. Laboratory workshop on water disposal and wastewater treatment [86]

As a rule, clean natural waters contain 0.01-0.1 mg/dm$^{(3)}$ of ammonium salt nitrogen.

Quantitative determination of nitrogen of ammonium salts.

Determination process. The volume of water for the study is taken taking into account the results of a qualitative sample with an approximate quantitative assessment. If the content of NH^{+4} in water is more than 0.5 mg/dm$^{(3)}$, the sample should be diluted.

To determine the amount of ammonium salts, the previously obtained solution is poured into a 10 mm FEC cuvette. Determination is carried out with a blue light filter with a wavelength of 400-425 nm. Distilled water is used as a control solution. The result is compared with the data in Table 9.

Table 9 - Ammonia content in water depending on the optical density of solutions

Optical density of solutions (by FEC)	Ammonia content mg/l	Optical density of solutions (by FEC)	Ammonia content mg/l
0,063	OD	1,130	1,8
0,070	0,2	0,138	2,0
0,080	0,4	0,146	2,2
0,085	0,6	0,153	2,4
0,092	0,8	0,161	2,6
0,100	1,0	0,168	2,8
0,108	1,2	0,176	h,o
0,115	1,4	0,183	3,2
0,123	1,6	0,191	3,4

Kalitsun V.I., Laskov Y.M. Laboratory workshop on water disposal and wastewater treatment [86]

Determination of nitrite nitrogen. The method is based on the interaction of nitrite and Griess reagent, which is a mixture of a-naphthylamine and sulfanilic acid in acetic acid, resulting in the formation of diazo compounds, coloured from pink to intensely red depending on the concentration of nitrite nitrogen.

Qualitative definition with approximate quantification.

Course of Determination.

In a flat bottom tube of colourless glass with a diameter of 13-14 mm add 10 $cm^{(3)}$ of test water and 0.5 $cm^{(3)}$ of Griess reagent. Place in a water bath at 50-60°C and heat for 10 min. Approximate nitrite content is determined according to Table 10.

The nitrite nitrogen content of pure natural waters is generally less than 0.005 mg/dm^3.

Table 10 - Approximate determination of nitrite nitrogen in water

Colouring on review		Nitrite nitrogen content mg/dm^3
On the side	From top to bottom	
No	No	Less than 0.001
Barely noticeable pink	Extremely faint pink	0,002
A very faint pink	Faint pink.	0,004
Faint pink.	Light pink	0,02
Light pink	Pink	0,04
Pink	Intense pink	0,07
Strong pink	Red	0,2
Red	Bright red	0,4

Kalitsun V.I., Laskov Y.M. Laboratory workshop on water disposal and wastewater treatment [86]

Quantitative determination of nitrite nitrogen in water.

Course of determination. If the nitrite content in water is more than 0.05-0.07 $mg/dm^{(3),}$ the sample should be diluted.

To 50 $cm^{(3)}$ of the test sample add 2 $cm^{(3)}$ of Griess's reagent, stir, place in a water bath at 50-60°C, after 10 min photometrically measured at 520 nm against distilled water to which Griess's reagent has been added.

The mass concentration of nitrite is found from the calibration graph.

To draw the calibration graph in measuring flasks with a capacity of 50 cm[(3) shall] be added 0, 0.1, 0.2, 0.5, 1.0, 2.0, 5.0, 10.0, 15.0 cm^3 of working standard solution and bring the volume to the mark with distilled water. Obtain solutions containing: 0; 0.002; 0.004; 0.01; 0.02; 0.04; 0.10; 0.20; 0.30 mg / dm^3 nitrite.

Then analysed and photometrically analysed as for the sample. According to the results obtained, a calibration graph is plotted with the mass concentrations of nitrite in milligrams in 1 dm^3 on the abscissa axis and the corresponding optical density values on the ordinate axis.

Nitrite content is determined according to the formula:

$$X = 50 \times C/V \ (mg/dm^3) \qquad (7)$$

Where:

C - nitrite content found from the calibration graph, mg/dm^3;

50 - volume of standard solution (dilution), cm^3;

V - volume of the sample taken for analysis, cm^3 [86-88].

2.3 Determination of the main indicators of soil contamination

The content of copper, chromium, zinc, lead and cadmium was determined in soil.

The main quality criteria are the values of maximum permissible concentrations (MPC) of pollutants in soil.

Determination of heavy metals in soil is carried out by atomic absorption spectrometry with flame and flameless atomisation. At present, for some elements, including copper, zinc, mercury, lead, etc., heavy metals are determined by atomic absorption spectrometry with flame and flameless atomisation.

Before the analysis, the soil from the bag is poured out on a flat surface, mixed well, spread in a layer no thicker than 1 cm and sampled from at least 5 places.

In order to convert the result of the analysis of air-dry soil sample to absolutely dry sample, the moisture content of the sample under study shall be determined during metrological evaluation of the methods.

Soil samples weighing 15 - 50 g are placed in pre-numbered, dried and weighed cups.

For clayey high humus soils with high moisture content 15 - 20 g is sufficient, for light soils with low moisture content - 40 - 50 g.

The cups with soil are weighed with an error of not more than 0.1 g. Cups with soil are opened and together with lids placed in a drying cabinet. Sandy soils are dried for 3 hours at 105°C, other soils - for 5 hours.

Subsequent drying for 1 h for sandy soils and 2 h for other soils. Plastered soils are dried for 8 hours. The subsequent drying is carried out for 2 h.

After each drying the cups with soil shall be covered with a lid, cooled in a calcium chloride desiccator and weighed with an error of not more than 0,1 g. Drying and weighing shall be stopped if the difference between re-weighing does not exceed 0.2 g. Soils with a high organic matter content may have a higher mass at repeated weighings than at previous weighings due to oxidation of organic matter during drying. In this case, the lowest mass should be used for calculations.

Soil moisture (W, %) is determined by the formula:

$$W = \frac{m_1 - m_0}{m_0 - m} \times 100\%. \qquad \dots\dots\dots(8)$$

Where:

m_1 - mass of wet soil with cup and lid, g;

m_0 - mass of dried soil with cup and lid, g;

m - mass of empty cup with lid, g.

Calculation is performed with an accuracy of 0.1%. Permissible differences of two parallel determinations are 10% of the arithmetic mean of repeated determinations.

Analyses

1. Chemical decomposition of soil samples for bulk determination of heavy metals. 10 g of air-dry soil, crushed and passed through a sieve with holes of 2 mm,

weighed on a technical scale, put the weight in a chemical beaker or conical flask with a capacity of 200 - 250 cc and add 50 cc of HNO (1:1). If the soil contains more than 5 % humus (according to Tyurin), preliminary dry ashing of the sample at 575 °C is recommended.

By rotating the flask gently stir the contents.

The beaker is covered with an hour glass and placed on a closed electric cooker, brought to a boil and boiled over low heat for 10 min.

Then 10 cc of concentrated hydrogen peroxide is added dropwise to the sample while stirring and again placed on an electric stove, brought to a boil and boiled for another 10 min.

After cooling to room temperature, the suspension is filtered through a folded filter "blue ribbon" into a measuring flask with a capacity of 100 cc, the filter with the precipitate is placed in a beaker, in which the rest of the soil was left. Pour 40 cc of 1 M nitric acid into the beaker and place it on the cooker, heat and boil for another 30 min.

After cooling down to room temperature, the liquid in the beaker is filtered into the same measuring flask. The precipitate on the filter is washed with hot nitric acid of concentration (HNO) = 1 mol/cubic dm and, after cooling, the volume of the filtrate in the measuring flask is brought to the mark with distilled water.

At the same time, a "blank" analysis is carried out, including all its stages except soil sampling.

2. Extraction of mobile forms of heavy metals from soils using acids.

The mobile acid-soluble forms of metals (Si, Zn, Ni, Co, Cd, Pb) are determined in fractions of 1 M HNO or 1 M HC1.

In recent years, these extractants have been successfully used for analyses of soils subjected to anthropogenic impacts. From heavily contaminated soils 1 M HNO extracts 90 - 95% of heavy metals from their gross content. Soil to solution ratio is 1:10, for peat soils - 1:20.

A 5 g soil sample (2.5 g for peat soils) is weighed with an accuracy of +/- 0.1 g and placed in a conical flask with a capacity of 200 - 300 cc, 50 cc of 1 M HNO is added to the sample (1 M HC1 can be used for extraction of Pb). The weight of soil should be increased up to 10 g when determining heavy metals at background level. In this case the ratio of soil and solution remains unchanged.

Shake the suspension on a rotator for 1 h or after 3-minute shaking insist for 24 hours. The flask is closed with a stopper (if rubber, it is necessary to wrap it with polyethylene film).

The extract is filtered through a dry white tape pleated filter prewashed with 1 M HNO. Before filtration the extract is mixed and transferred to the filter as completely as possible. In the filtrate heavy metals are determined on the atomic absorption spectrophotometer in acetylene-air flame. If the filtrates are turbid, they are returned to the filters. At the same time the blank analysis is carried out, including all stages of its determination, except for sampling.

The content of metals in the studied soil samples is calculated by formula (9):

$$x = \frac{V \times (A_1 - A_0)}{m}, \qquad (9)$$

Where:

x - mass fraction of the determined metal in air-dry soil sample, mln^{-1}, (mg/kg);

Ai - metal concentration in the studied acid (buffer) soil extract, found from the calibration graph, mg/cubic dm;

Ao - metal concentration in the control sample, found from the calibration graph, mg/cc dm;

V - volume of the tested solution, cc. cm;

m - mass of air-dry soil sample, g [90].

CHAPTER 3

INTEGRATED ECOLOGICAL AND HYGIENIC ASSESSMENT OF THE STATE OF THE ENVIRONMENT IN THE STUDIED REGIONS OF AKMOLINSK OBLAST

3.1 Ecological and hygienic assessment of atmospheric air condition in districts of Akmola region with different technogenic load

It is known that atmospheric air pollution by a complex of various harmful substances leads to unfavourable shifts in the health of the population and, first of all, children. Due to the fact that the human body is much more sensitive to the penetration of toxic substances through the lungs, the state of the atmosphere is an important problem that requires careful study [34].

Analysis of the state of atmospheric air pollution in Akmola region as a whole consists of emissions of harmful substances from stationary sources and motor transport.

Gross emissions of pollutants into the atmosphere from all sources in 2014 amounted to 338789.94 tonnes. At the same time, gross atmospheric emission from stationary sources is 209375.46 tonnes, from mobile sources - 129414.48 tonnes (Figure 3).

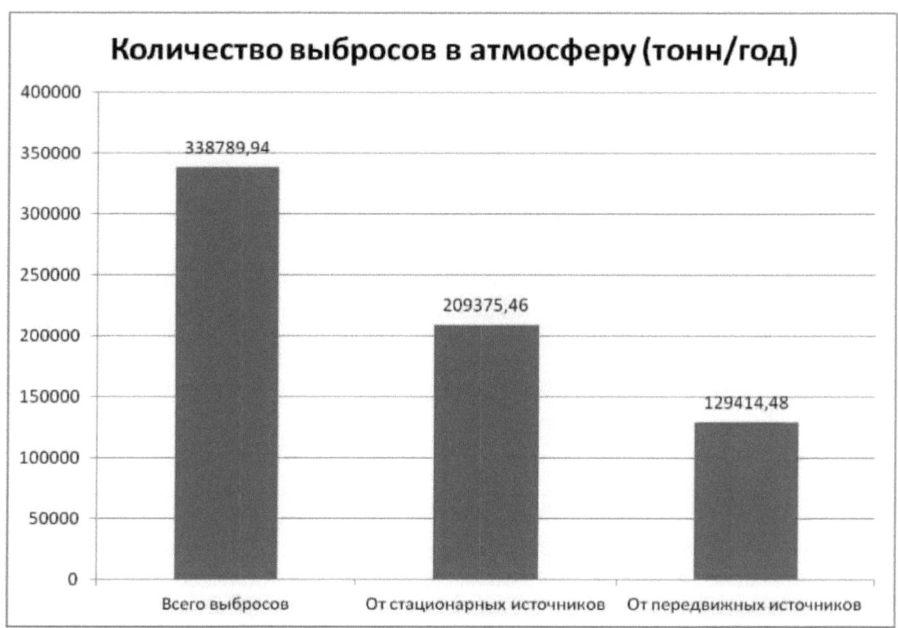

Figure 3 - Emissions of pollutants into the atmospheric air in Akmola region in 2014 (tonnes/year) [developed by the author] As follows from the diagram, emissions from stationary sources exceed emissions from motor transport almost 1.6 times.

On the territory of Akmola region the main sources of atmospheric air pollution are stationary sources of mining, industrial,   fuel and energy, transport and road, agricultural and other enterprises (Figure 4).

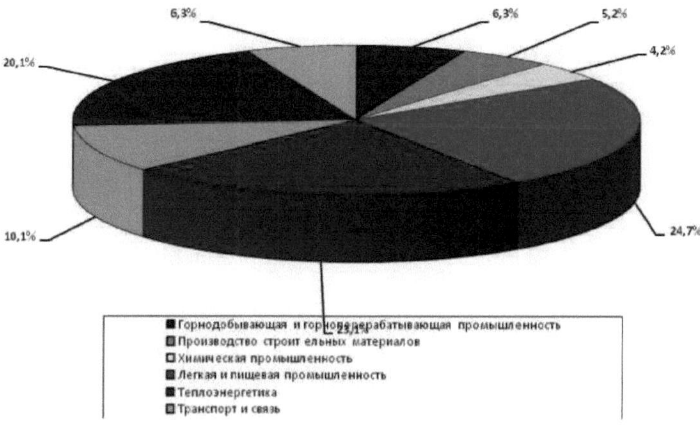

Figure 4. - Contribution of production sectors to atmospheric air pollution in Akmola Oblast

Emissions from stationary sources are dominated by carbon monoxide, nitrogen dioxide, sulphur dioxide, phenol, formaldehyde and dust (Figure 5).

Figure 5. - Structure of emissions to atmospheric air from stationary sources, 2014

The industrial complex of the region, which accounts for about 18.3% of the gross regional product, is represented mainly by mining, machine building, non-ferrous metallurgy, chemical and food industries, and construction industry.

The main industrial enterprises of the region are:

- JSC MMC Kazakhaltyn and Altyntau Kokshetau LLP - gold production;

- JSC Tynys - production of aircraft assemblies and units, fire-fighting equipment, gas shut-off valves, medical and weight-measuring equipment, polyethylene pipes;

- Stepnogorsk Bearing Plant JSC - production of bearings for railway rolling stock. In addition to domestic consumption, a significant part of production is supplied to Russia;

- Baiterek-A JSC - overhaul of electric locomotives;

- Stepnogorsk Mining and Chemical Combine LLP - production of uranium, molybdenum concentrate;

- Kokshetau Mineral Waters JSC - production of liqueur and vodka products, soft drinks and mineral water;

- JSC KAMAZ-Engineering - organisation of production of automotive

48

equipment;

- Orken-Atansor LLP - iron ore mining;

- KazShpal JSC - production of reinforced concrete structures.

There is a developed network of enterprises for processing of agricultural raw materials (meat processing plants, butter factories, mills, bread receiving enterprises, beverage production enterprises), light industry (clothing and textile production).

The largest stationary sources of atmospheric pollution in Akmola region are Stepnogorskaya TPP of "Jet-7" LLP and SCP on PCW "District Boiler House No.2" of Kokshetau city.

The main contribution to atmospheric air pollution by gross emissions belongs to Stepnogorsk CHPP, on average 32.8% of the total industrial emissions. The CHPP emits 243.5 tonnes of sulphur dioxide, 121.4 tonnes of carbon oxide, 324.8 tonnes of nitrogen dioxide and other chemicals per year.

Table 11 shows that the districts of the Oblast have their own specificity of air pollution associated with the nature of industries located in them. Thus, in Zheleznodorozhny and Severny districts the highest indicators of air pollution by industrial enterprises (P = 6.8 and 5.5) are observed, while in Zavodskiy district it is much lower.

For better clarity, let us present tabular data in the form of diagrams (Figure 6).

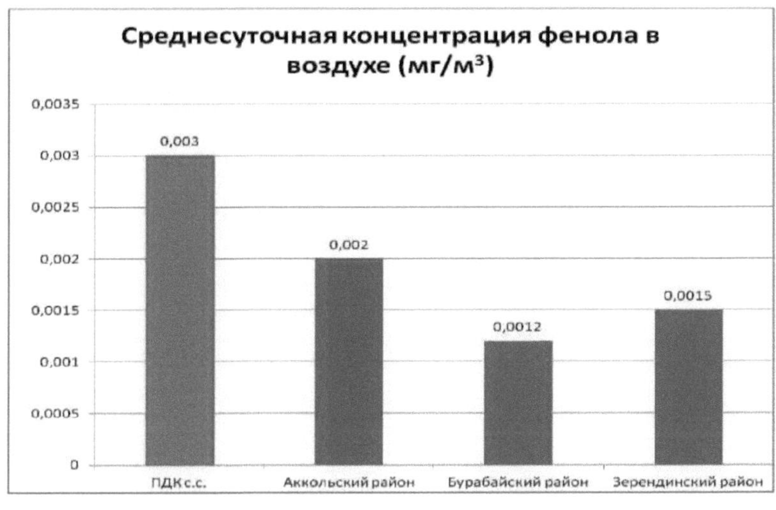

Figure 6. - Atmospheric air pollution by phenol in the studied districts of Akmola region

As can be seen in the diagram, in the studied districts of Akmola region average daily concentration of phenol in the atmospheric air does not exceed the maximum permissible level, but in Akkol district its concentration is the highest, which is apparently due to the presence of chemical industry in this region, in the territory of the district there are 10 enterprises whose emissions exceed 50 tonnes/year. The hygienic importance of phenol is very high. Phenol itself is a very toxic compound, has a pronounced mutagenic effect, and under certain climatic and weather conditions can form more dangerous compounds [19]. Phenol is readily absorbed through the skin and gastrointestinal tract, and phenol vapours are readily absorbed through the lungs. The toxic effects of phenol are directly related to the concentration of free phenol in the blood. Phenol is a common protoplasmic poison and is toxic to all cells. Phenol belongs to the group of substances of hazard class 2 in terms of impact on public health [22].

Next, consider the air content of the equally harmful compound formaldehyde (Figure 7).

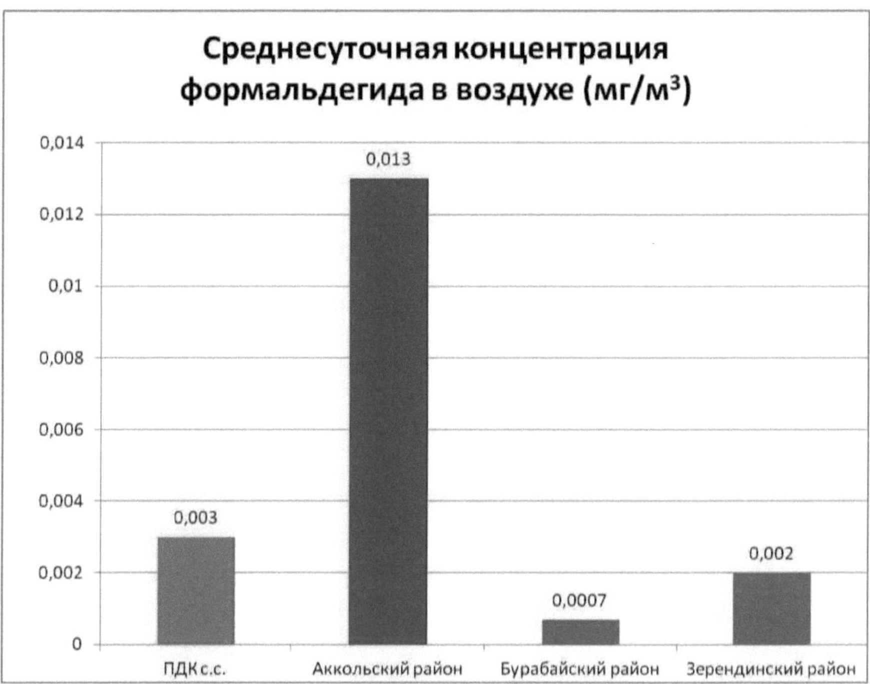

Figure 7. - Atmospheric air pollution by formaldehyde in the studied districts of Akmola region

The diagram shows that in Akkol district the average daily concentration of formaldehyde is more than four times higher than the maximum permissible level. In Burabay and Zerendinsky districts this indicator is within the norm. Formaldehyde is classified as a toxic and hazardous substance of the 2nd class of hazard. At elevated concentrations, formaldehyde has a variety of toxic effects: irritates mucous membranes of the upper respiratory tract, throat, eyes, causes nausea and headache [15]. In addition to its general toxic effect, this compound is a carcinogen [19, 22].

Figure 8 is a diagram showing the average daily concentration of sulphur dioxide gas.

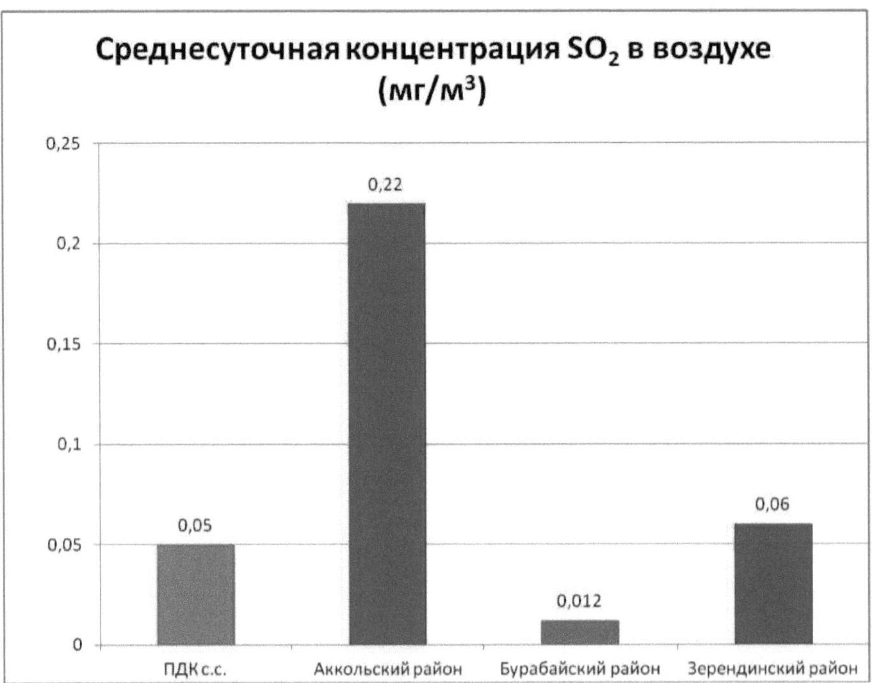

Figure 8. - Atmospheric air pollution by sulphur dioxide in the studied districts of Akmola region

As follows from the diagram, in Akkol district the content of sulphur dioxide in the air exceeds the maximum permissible concentration by 4.4 times, also insignificant excess of MPC was recorded in Zerendinsky district, in Burabay district the concentration of SO2 is within the norm. Sulphur dioxide gas is classified as a hazard class 3 substance. Poisonous effect of sulphur dioxide is expressed in irritating effect on mucous membranes of upper respiratory tract and eyes. In chronic poisoning, catarrhs of the upper respiratory tract, conjunctivitis, severe bronchitis and other damage to internal organs develop [18].

Figure 9 presents a diagram showing the average daily concentration of carbon monoxide.

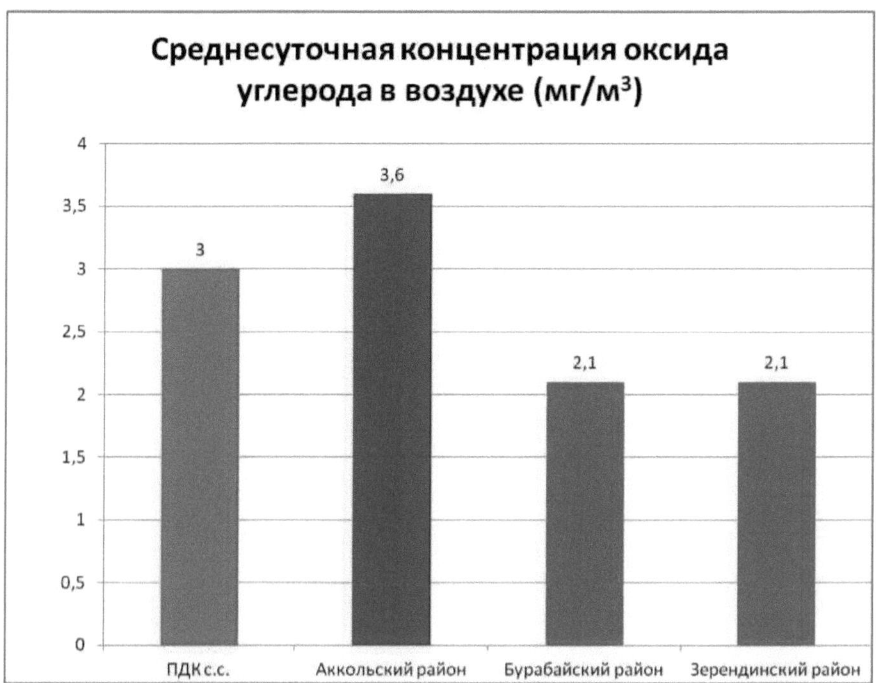

Figure 9. - Atmospheric air pollution by carbon monoxide in the studied districts of Akmola region

The diagram shows that in Akkol district there is a slight excess (0.8 times) of carbon monoxide concentration, in Burabay and Zerendinsky districts this indicator is within the permissible level.

Carbon monoxide belongs to the group of substances with hazard class 4. The impact of high concentrations of carbon monoxide on the human body is manifested as follows: the gas prevents the absorption of oxygen by the blood, which impairs thinking ability, slows reflexes, causes drowsiness and may be the cause of unconsciousness and death [18,20].

Figure 10 presents a diagram showing the average daily
the concentration of nitrogen dioxide in the air.

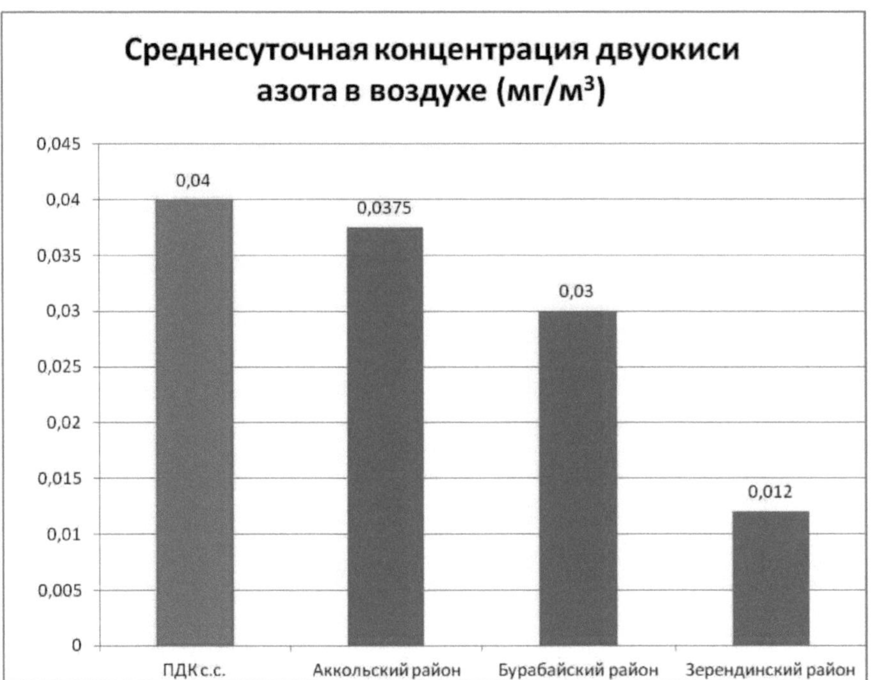

Figure 10. - Atmospheric air pollution by NO2 in the studied districts of Akmola region

According to the data presented in the studied districts of Akmola region, the average daily concentration of nitrogen dioxide in the atmospheric air does not exceed the maximum permissible level, but in Akkol district its concentration is the highest.

Nitrogen dioxide belongs to the group of substances of hazard class 3. The functional effect caused by nitrogen dioxide is increased airway resistance. In other words, NO2 causes an increase in respiratory effort. The pathological effects are that NO2 makes a person more susceptible to pathogens that cause respiratory disease. People exposed to high concentrations of nitrogen dioxide are more likely to experience upper respiratory tract catarrh, bronchitis, croup, and pneumonia. In addition, nitrogen dioxide itself can cause respiratory disease [16, 22].

Thus, we can conclude that the most unfavourable among the three studied districts of Akmola region is Akkol district, here exceeding the average daily concentrations of formaldehyde, sulphur dioxide and carbon monoxide is observed. In

Burabay district the situation is the most favourable, no exceedance of maximum permissible level is observed for any of the substances considered. In Zerendinsky district there is a small excess of MPC s.s. for sulphur dioxide, in general, the environmental situation is favourable.

In addition to industrial sources of pollution, motor transport has a negative impact on air quality, emissions from which account for 38.1% of all pollutants in the Oblast.

It should be noted that, unlike industrial sources, motor transport pollutes the surface layer, which can pose a great danger to the environment and human health.

Consequently, pollution of the air basin of the region from stationary and mobile sources creates an unfavourable environmental situation, which can have a negative impact on the protective functions of the organism and contribute to the development of various diseases.

3.2   Ecological and hygienic assessment of surface water quality in some districts of Akmola region with different anthropogenic load

The problem of providing the population of Akmola oblast with ecologically good quality drinking water is one of the most socially significant, as it directly affects human health.

Let's consider the state of water resources in the study areas.

In Aksu district, the hydrological network is represented by the rivers Tolkara, Aksuat, Koluton, Stepnaya, Tasmola, Aschily-Airyk, Aksu and lakes, the largest of which are Zharlykol, Itemgen, Shortankol, Balyktykol, Akkol, Zharsar. In addition to large lakes, the territory of the district is represented by many small lakes with a small area of water mirror[13].

Water supply of settlements and industrial enterprises is carried out from underground and surface water sources.

Centralised wastewater disposal is carried out only in the city of Akkol. Wastewater is discharged to filtration fields. Sewerage networks were put into operation in 1979 and currently have a poor technical condition. The volume of wastewater generated currently exceeds the designed volumes, and in this regard, emergency situations often occur at certain sections of the sewerage collector, resulting in the discharge of wastewater to the terrain. In other settlements wastewater is discharged to the terrain in an unorganised manner.

The largest water basins of Zerendinskiy district are Lake Zerendinskoye (10.7 km^2 in area) and the Chaglinka River, as well as many freshwater and saltwater lakes, small rivers, underground springs and reservoirs. The main sources of pollution of water bodies are wastes from agricultural, industrial and municipal activities, sewage.

There are 3 reservoirs on the territory of Zerendinsky district: Chaglinskoe (area - 660 ha, volume - 28.0 million $m^{3)}$ on the river Chaglinka (river length - 145 km), on the river Tepic butax. Podlesnoye, on the river Kylshakty in Akkol rural district. In addition, there are 38 lakes with a total area of 7037.4 ha. The lakes differ both in size and in morphometric indicators of water (area, depth, water mass volume, overgrowth, salinity). The largest lakes in the district are: Zerenda, Aidabol, Kumdykol. The average depth of lakes varies from 1 metre (Kurkol, Sualma) to 5 metres (Zerenda). Most of the lakes are fresh water bodies. Such large lakes as Zerendinskoye, Aidabol and Korovye are located in the territory of the GN1P1 "Kokshetau". Zerendinskiy district can be characterised as water-supplied in terms of the number of reservoirs, the district is comparatively favourable in terms of the quality of consumed water[28, 29].

Most of the water is taken from open sources - Lake Zerendinskoye, rivers Teres, Butak, Argaly, Chaglinka, Aidabulka. Smaller part of water is taken from underground sources. Only one enterprise, JSC "Aidabul distillery", operates on recycled water supply. The above-mentioned rivers of the district are severely depleted and silted up, it is connected with construction of non-designed dams and crossings.

Water supply to the population and economic entities of the district is carried out through a system of 152.5 km of water distribution networks (which require major repairs), 57 wells, 68 mine wells for public use, artificial and natural reservoirs, and individual sector wells. In total, there are 24 water users in the district. Total water withdrawal for 2014 was 1.25 million metres^3, including 0.532 million metres^3 of surface water.

The length of sewerage networks in the district is 51.5 km. Wastes and sewage accumulate in cesspools and serve as a source of ground and surface water pollution. There is no storm water drainage system in the district centre, surface polluted water from the territories is washed into Lake Zerendinskoye, worsening water quality. Wastewater from the district centre and recreation area is transported by motor transport to the central district landfill.

Unsatisfactory sanitary condition has settlements located in the water protection zone of the Chaglinka River, on its catchment area. As a result of untimely cleaning of settlements from rubbish and manure, pollutants are washed away and subsequently get into the Chaglinka reservoir, which is the source of water supply for Kokshetau city.

The hydrographic network of Burabay district is represented by lakes. The largest one is "Bolshoye Chebachee". The total catchment area is 150 km2. The area of the water mirror is 22 km^2. Average water volume - 240 million metres^3. Mineralisation - 4.0 - 6.0 g/l. Hardness 3-5 mg/eq (moderately hard). Chemical composition of water is hydrocarbonate with predominance of sodium and significant content of magnesium ions. Pollution index - moderately polluted, class - 3.

Lake Shchuchye. Total catchment area - 64.4 km2, average water mirror area - 18 km2. Average water volume - 225 mln m^3, hardness 2.0 - 2.5 mg/eq (soft). Chemical composition of water - hydrocarbonate-calcium. Pollution index - moderately polluted, class - 3. Water from the lake is used for domestic drinking water supply of Shchuchinsk.

Borovoye Lake. Total catchment area - 164 km2, average water mirror area - 10 km2, average water volume - 29 million m^3, mineralisation - 1.01.5 g/l, hardness 1.0-1.5 mg/eq (very soft). Chemical composition of water - hydrocarbonate-calcium. Pollution index - moderately polluted, class - 3.

Lake "Maloe Chebachee". Total catchment area - 139 km2, average water mirror area - 21 km2, average water volume - 131 million m^3. Mineralisation - 2.3-2.7 g/l, hardness 15-25 mg/eq (very hard).
Chemical composition of water - sodium chloride with significant content of magnesium. Pollution index - dirty, class - 5.

Kotyrkol Lake. Total catchment area - 29.9 km^2, average water mirror area - 4.5 km^2. Water composition is sodium hydrocarbonate, hardness - 3.8 mg/eq (moderately hard), pollution index - very dirty, class - 5.

Other lakes of the district have small water mirror area and different chemical composition. All the above lakes have negative water balance, water inflow equals summer evaporation. On Lake Bolshoye Chebachee, for example, several islands and peninsulas have even formed in recent years. As a result of shoaling, water temperature and salinity increased, which led to intensive development of algae, which later, when they die off, led to intensive siltation of lakes and deterioration of water quality. Another factor increasing annually the thickness of silt deposits is the area washout of humus from arable lands located in the catchment area of the lakes[23, 24].

There are practically no rivers in the rayon; the existing rivers dry up in summer. There are 87 water users on the territory of the rayon. Treatment facilities of Shchuchinsk city are on the balance of the municipal water supply company, commissioning date - 1938, capacity of treatment facilities - 10.5 thousand m^3/day.

Water intake is carried out from Lake Shchuchye. Water intake facilities are located on the shore of the lake. Given the drop in the lake level, there is a need to move the pumping station closer to the water. Chlorination of water is done in a rudimentary manner and often disinfection is not carried out due to lack of funds to purchase disinfectants. The water distribution networks require major repairs.

Water supply in the district farms is extremely unsatisfactory. Water distribution networks are not washed and disinfected in time. Residents of Yurievka, Mezgilsor, Karabaur, Aigabak, Savinka villages are forced to use water from open water bodies and dig wells. And since the groundwater table in Burabay rayon is high enough, people, digging wells 2-3 metres deep, simply drink groundwater, i.e. topsoil water. Water pipelines in rural areas are in a neglected sanitary and technical condition [22].

Only Shchuchinsk is a sewered settlement. Due to the lack of sewerage network in Borovoye settlement and health resorts located in Borovoye, sewage is discharged into cesspools with further transportation and accumulation at the settlement dump.

Dorofeevka village, located on the shore of the Maloye Chebachee lake, is the main polluter of this lake. The villagers store manure and slag near their houses and do not remove it regularly. During flash floods, all the sewage from the village is washed away and freely flows into the lake[25].

A number of sanatoriums located on the shore of Lake Shchuchye and having liquid fuel boilers allow spills of oil products in large volumes.

3.3 Ecological and hygienic assessment of soil quality in some districts of Akmola region with different anthropogenic load

Soil, being a natural environment, is in close interaction with atmospheric air, surface and underground waters, and actively exchanges substances and energy.

The main quality criteria are the values of maximum permissible concentrations (MPC) of pollutants in soil. The results of analyses of soil samples are presented in Table 12.

Table 12 - Main indicators of soil pollution by districts of Akmola region for 2014

Name impurities	MPC (mg/kg)	Akkolsky (mg/kg)	Burabay (mg/kg)	Zerendinsky (mg/kg)
Copper	3,0	4,6	2,8	3,9
Chrome	6,0	5,1	4,0	4,2
Zinc	23,0	21,6	15,3	17,1
Lead	32,0	27,0	П,2	19,6
Cadmium	0,5	0,3	0,1	0,1

For better clarity, let us present tabular data in the form of diagrams (Figure 11).

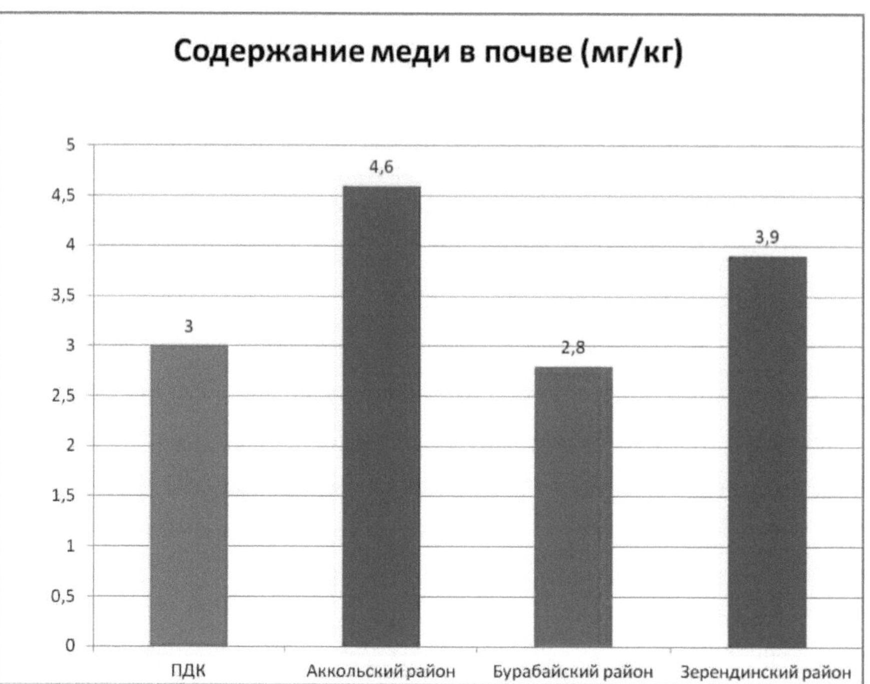

Figure 11. - Soil contamination with copper in the studied districts of Akmola region

As can be seen in the diagram, in Akkol district there is an excess of copper content in soil samples in 1.5 times, a small excess is in Zerendinsky district. In Burabay district copper content is within the norm [22]. The effect of heavy metals on living organisms is often hidden, but they are transmitted along trophic chains with a pronounced cumulative effect, so manifestations of toxicity can occur unexpectedly at individual levels of trophic chains. Excess copper has harmful effects on the organism of warm-blooded organisms. Copper belongs to the group of highly toxic metals capable of causing acute poisoning of humans and animals and having a wide range of toxic effects with a variety of clinical manifestations. Copper belongs to the 2nd class of hazard [23].

Next, consider the chromium content of the soil (Figure 12).

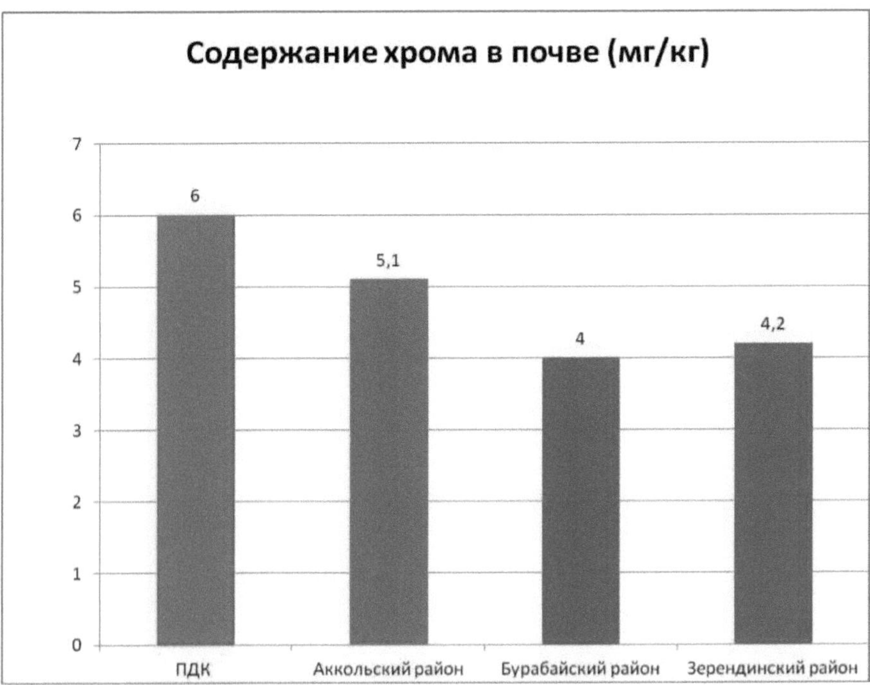

Figure 12. - Soil contamination with chromium in the studied districts of Akmola oblast

As can be seen in the diagram, in the studied districts of Akmola region the concentration of chromium in soil does not exceed the maximum permissible level, but in Akkol district its concentration is the highest. Toxicity of chromium compound is in direct dependence on its valence: chromium (VI) compounds are the most toxic, chromium (Sh) compounds are highly toxic, metallic chromium and its compounds (II) are less toxic.

Regardless of the route of entry, the kidneys are affected first. Liver and pancreas functions are also affected. Chromium has a carcinogenic effect, affects the CNS, has a damaging effect on reproductive function. It is classified as a hazard class 1 substance [19].

Figure 13 is a diagram showing the zinc content of the soil.

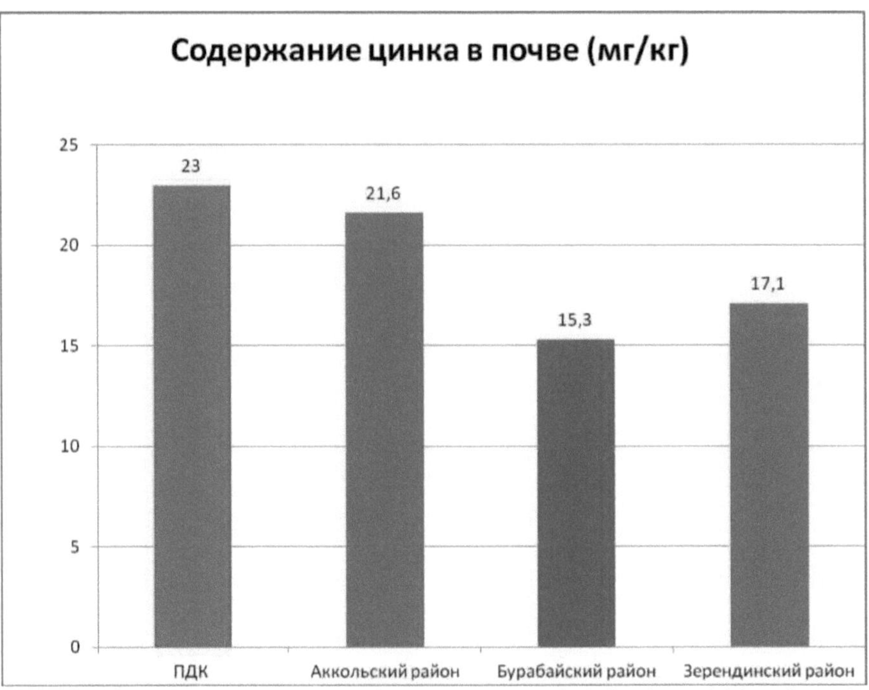

Figure 13. - Soil contamination with zinc in the studied districts of Akmola region

As follows from the diagram, in Akkol district zinc content in soil is higher compared to other districts, but, however, does not exceed the maximum permissible concentration. Zinc is a trace element necessary for normal functioning of human organism in small doses. It is a member of 40 metalloenzymes that play an important role in the metabolism of nucleic acids and protein synthesis. Metallic zinc has little toxicity. Zinc phosphide and zinc oxide are poisonous. The ingestion of soluble zinc salts leads to digestive disorders, irritation of mucous membranes. Zinc belongs to the substances of the 2nd class of danger. [18].

Next, Figure 14 presents a diagram showing the lead content in the soil.

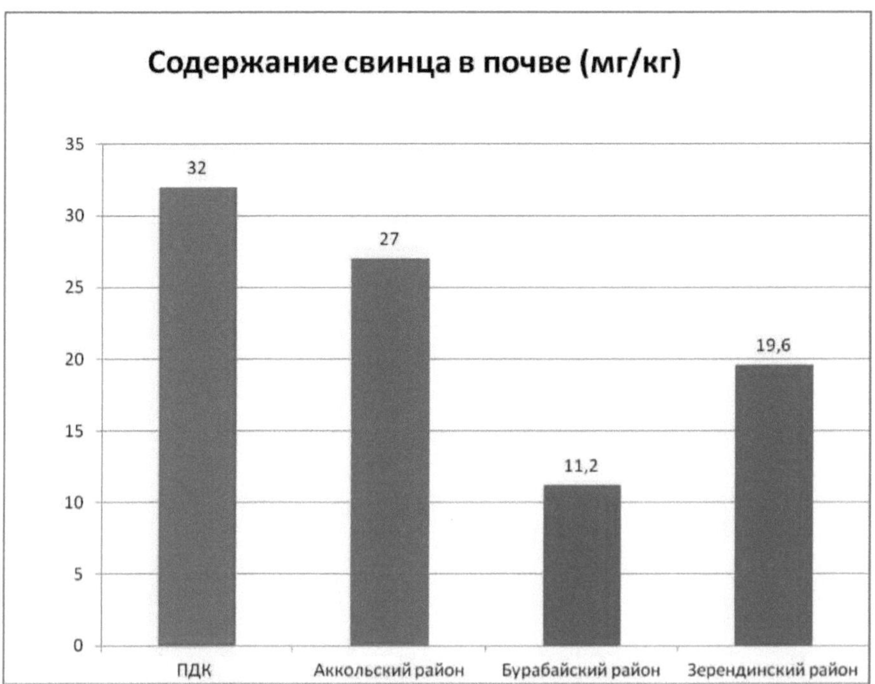

Figure 14. - Soil contamination with lead in the studied districts of Akmola region

The diagram shows that in all investigated districts of Akmola region the content of lead in soil is within the permissible level.

Lead by its effect on the human body belongs to substances of hazard class 1. Lead affects the human nervous system, which leads to a decrease in intelligence, causes changes in physical activity, hearing coordination, affects the cardiovascular system, leading to heart disease. This has a negative impact on the health of the population and primarily children, who are most susceptible to lead poisoning. [18, 20]. Figure 15 presents a diagram reflecting the content of no less dangerous element cadmium in the soil.

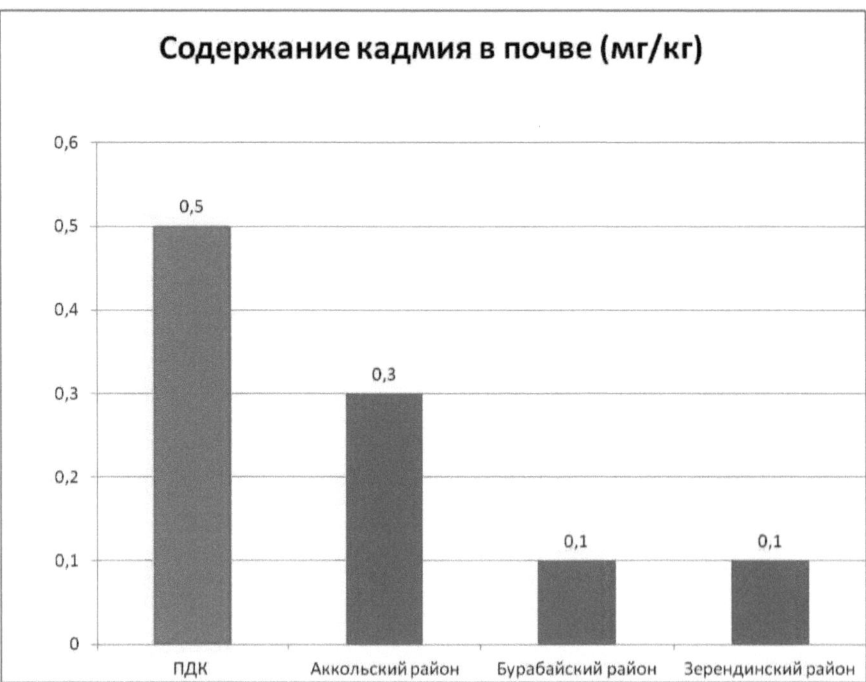

Figure 15. - Soil contamination with cadmium in the studied districts of Akmola region

According to the data presented in the studied districts of Akmola region, cadmium content in soil samples does not exceed the maximum permissible level, but in Akkol district its concentration is the highest.

Cadmium vapour, all of its compounds are toxic due to its ability to bind sulphur-containing enzymes and amino acids. Cadmium can raise blood pressure. It has carcinogenic effects. Cadmium accumulates in kidneys, during human life its content can increase 100-1000 times. The hazard class of the substance is 1. [16, 22].

As the chemical analysis of soil of the study areas showed, the content of such substances as cadmium, zinc, lead, chromium do not exceed MPC. However, 1.5 times of copper exceeds MAC in the soil of Akkol district and 0.8 times in the soil of Zerendinsky district.

Thus, the conducted studies of water and soil have shown that there is no significant difference in the content of chemical substances between districts, however, the districts of the region have their own specificity of air pollution, depending on the nature of industries located in them.

3.4   Characteristics of medical and demographic condition of the population of Akmola region

3.4.1 Physical development of children and adolescents living in areas of the region with different anthropogenic load

The physical development of children and adolescents is one of the most important and highly sensitive indicators of the health of the younger generation, which depends in no small measure on the ecological well-being of the environment.

Analysing body length indicators by age, we found that on average 51.3% of children were within the normal range. The number of stunted children (with reduced and low body length) is 17.5% and 4.7%, with 3.3% of children having severe growth retardation. The number of tall children is 23.2 per cent.

Analysis of body weight by age showed that children with normal body weight indicators (25-75 centiles) predominate among the surveyed children in schools, their number averaging 58.6 per cent. Reduced and low body weight is observed in an average of 20 per cent and 8.7 per cent of children, respectively, and 4.3 per cent of children have a pronounced body weight deficit (less than 3 centiles). At the same time, 8.4 per cent of children are overweight (more than 97 centiles).

The assessment of the harmoniousness of children's physical development has shown that normal physical development is observed in the majority of children, averaging 57.5 per cent. Deviations in physical development are mainly represented by the group of children with reduced body length at normal, reduced and low body weight values, which accounts for 23.7 per cent of children, and with reduced body weight at normal length values - 11.2 per cent. The share of children with increased and high body weight at normal body length values is 5.5 per cent. Low body weight with normal body length is noted in 2.1 per cent of children. However, when comparing the physical development indicators of school children from different districts, no reliable differences were revealed, i.e. physical development was almost the same in different districts of the Oblast.

The results of the study of adolescents' physical development showed that 53.4 per cent of adolescents have normal body length in relation to age. On average, 18.9 per cent of adolescents have reduced and low growth, 2.6 per cent of them have more pronounced growth retardation, and 25.1 per cent of adolescents have growth exceeding age norms.

61.5 per cent of adolescents have a normal body weight, 16.7 per cent have a reduced body weight, and 14.5 per cent of adolescents have a body weight deficit, while 2.9 per cent have a significant deficit. Excess body weight is observed in 4.4 per cent of adolescents.

The distribution of adolescents by physical development groups showed that 57.9 per cent of students belonged to the group with normal physical development, while 42.1 per cent of adolescents had certain developmental abnormalities. Thus, 19.7 per cent of adolescents have reduced body length with normal, increased and high body weight values. The number of children with reduced body weight with normal body length values is 11.6 per cent, and with increased and high body weight with normal body length values is 6.3 per cent.

3.4.2 Characteristics of morbidity of the population living in areas of the region with different technogenic load.

An analysis of the structure of morbidity among the child population in 2014 showed that the first place is occupied by diseases of the respiratory organs, which account for 70 per cent of the total morbidity of the child population. The second place is occupied by accidents, injuries and poisonings - 4 per cent, the third and fourth places are occupied by diseases of digestive organs - 3.8 per cent and of the musculoskeletal system - 2.6 per cent, the fifth place is occupied by diseases of the nervous system and sense organs - 2.4 per cent.

However, in Zerendinsky district, diseases of the digestive system ranked second and accounted for 6% of the total morbidity, while diseases of the endocrine system ranked third (4.7%). In this district, congenital anomalies were 2-3 times higher and circulatory system diseases were 4.3 times higher (p<0.01). It should also be noted that diseases of the respiratory system are the most common in Akkol district and account for 75% of the total morbidity, the second and third places are occupied by diseases of the nervous system-2.85% and hematopoietic organs-2.71%. The incidence of neoplasms is 1.2 times (p<0.05) higher in Zerendinsky district and is 10.3 per 1000 population.

In the structure of adolescent morbidity, as in children, respiratory diseases predominate and account for 62 per cent of the total morbidity of adolescents, while diseases of the endocrine system, nervous system and genitourinary system are more frequently registered among adolescents. Diseases of the nervous system (8.87 per cent) and injuries, poisonings and accidents (8.82 per cent) occupy the second and third ranks among adolescent diseases, followed by diseases of the musculoskeletal system (7.2 per cent), digestive organs (6.4 per cent), genitourinary system (4.16 per cent) and endocrine system (1.83 per cent).

The study of morbidity by districts of the region showed that the level of general morbidity among adolescents in Akkol district, compared to others, is higher and amounts to 1224.15 per 1000, here the most common are diseases of respiratory organs, endocrine system, nervous system. The incidence of digestive diseases in different districts of the region among adolescents is approximately at the same level.

A study of the structure of adult morbidity revealed that respiratory diseases (27.6 per cent) and diseases of the circulatory system (15.53 per cent) occupy the leading places, followed by diseases of the nervous system (8.42 per cent), diseases of the musculoskeletal system (7.48 per cent) and diseases of the digestive system (5.87 per cent).

Thus, diseases of respiratory organs, neoplasms, diseases of blood and haematopoietic organs, digestion and other organs are more common in Akkol and Zerenda districts, where the degree of atmospheric air pollution is higher.

Thus, the analysis of morbidity of the population of different age groups showed some differences in nosoforms in different districts of the region, which, apparently, are caused not only by differences in socio-economic terms, but to some extent depend on the environmental conditions of their residence.

When assessing correlations between morbidity and environmental factors, positive correlations were obtained between specific air pollutants and the level of some diseases, for example, between sulphur dioxide and morbidity of children with neoplasms (r=0.63), diseases of the endocrine system (r=0.7), circulatory system (r=0.68), with diseases of blood and hematopoietic organs in adolescents (r=0.41).In addition, there is a positive correlation of carbon monoxide with diseases of the circulatory system (r=0.4) and anomalies (r=0.44) in children and diseases of the endocrine system (r=0.47) in adolescents. The incidence of diseases of the nervous system and sensory organs in children and adults has a pronounced correlation with nitrogen dioxide (r=0.47 and r=0.42). As the intensity of atmospheric air pollution by phenol increases, the incidence of diseases of nervous system and sense organs is more frequently registered (r=0.62 in children, r=0.5 in adolescents, r=0.4 in adults), abnormalities in children (r=0.46), in adults - neoplasms and diseases of circulatory system (r=0.4). In areas with high total concentrations of formaldehyde in the air there is an increase in the incidence of neoplasms in adolescents (r=0.42), in children and adults diseases of blood and hematopoietic organs (r=0.6 and r=0.5), in adolescents and adults diseases of digestive organs (r=0.4 and r=0.5). It was found that diseases of respiratory organs with general atmospheric air pollution have a weak positive relationship, which increases (r=0.4) with increasing concentration of formaldehyde in the air.

Thus, the obtained results of the study allowed us to identify priority environmental factors affecting the health of children, adolescents and adults in the region: increased air pollution contributes to the increase in morbidity. However, the obtained results are not unambiguous and require further more comprehensive studies of environmental factors.

3.5   Practical recommendations to reduce the impact of anthropogenic load on public health

The state of health of the population of Akmola region, the sanitary and epidemiological situation and the development of the health sector in the last decade were characterised by both positive and negative indicators [91].

The solution of environmental problems is possible only on the basis of a comprehensive approach, which includes socio-political, organisational and legal, economic, technological, educational, propaganda and some other aspects [92]. All of them can be combined within the framework of two general directions of solving environmental problems: greening of society and optimisation of nature management [93].

Ecologisation of society means the formation of such a system of attitudes, views, values, in which understanding of environmental problems becomes an urgent, natural need of each citizen individually and society as a whole [94, 95]. This is possible on the basis of broad environmental education and upbringing, on the basis of formation of ecological thinking [96].

An ecologically educated and brought up person has a sense of internal prohibition of any actions that can damage nature and thus society [97]. Only such a person can solve any tasks, including technological ones, related to the elimination of ecological crisis situations, the restoration of ecological stability and the normalisation of human-nature relations [98].

Ecologisation of society is environmentally sound organisational and legal decisions taken by representative, executive and judicial authorities [99]. It is full informing of the population through mass media about the environmental situation and the degree of its impact on human health [100].

Ecologisation of the whole society, when every citizen is ecologically brought up, ecologically educated and has ecological thinking, will allow to fully solve the problems of rational use of natural resources [101].

Optimisation of nature management is a scientific and technical solution to environmental problems, providing a way out of ecological crisis situations and normalisation of interaction between society and nature [102]. Optimisation of environmental management includes a number of measures for the rational use of natural resources:

- development of resource-saving and other environmental technologies;

- scientific substantiation of economic projects, including their environmental expertise, forecasting and environmental monitoring;

- purposeful regulation of the structure of natural complexes and protection (conservation) of some of them.

The main objective of solving environmental problems is to create conditions for stabilising and improving the quality of the environment, favourable for the population. This goal will be achieved by solving the following tasks:

- reducing the level of anthropogenic impact on the natural environment;

- reduction of pollutant emissions into the air basin (motor transport, heat and power generation, industrial enterprises);

- protection and rational use of water resources;

- improving the system of collection, storage, disposal and recycling of industrial and household waste;

- improving environmental culture and education of the population;

- environmental management system 180 14001:2004 [103].

The main directions for obtaining the result of the set goal are:

- carrying out measures to reduce emissions of pollutants into the air basin;

- prevention of emergency and environmentally hazardous situations at sewerage facilities;

- Exclusion of pollutants entering surface sources;

- providing the population with quality drinking water;

- improving the system of collection, storage, disposal and recycling of industrial and household waste;

- improving environmental culture and education of the population;

- implementation of the environmental management system 150 14001:2004 at the enterprises of the region [104].

Reduction of pollutant emissions into the air basin.

In order to reduce emissions of pollutants into the air basin, it is necessary to:

- to continue work on the transition of motor vehicles to alternative fuels (introduction of gas-cylinder equipment at car enterprises of the city and the region);

- to create protective zones of green areas on busy motorways of the city and the region in order to suppress dust and reduce the concentration of harmful emissions into the air from mobile sources;

- purchase and installation of dust and gas purification equipment at the facilities [105].

Protection and rational use of water resources.

Prospects for the protection and rational use of water resources are related to the prevention of emergency and environmentally dangerous situations at sewerage facilities, the exclusion of pollutants in surface sources, and the provision of the population with quality drinking water [106].

In this regard, it is assumed that:

- overhaul of treatment facilities, sewerage networks;

- overhaul of the city sewerage system collector in order to prevent sewage infiltration;

- surface and groundwater protection measures, installation of water protection signs;

- carrying out scientific and research works in terms of studying the problems of waterlogging of settlements, as well as disturbance of ecosystems of surface water sources.

Improvement of the system of collection, storage, disposal and recycling of industrial and household waste.

The following measures are expected to be implemented:

- design and construction of landfills for storage and disposal of toxic waste;

- arrangement of MSW dumps, cleaning, elimination and removal of illegal dumps in rural settlements;

- development of solid waste management programmes taking into account landfill reconstruction;

- In rural settlements it is necessary to organise wastewater removal to specially equipped sites for its reception.

- reclamation of disturbed land, restoration and enhancement of fertility and other useful properties of land and its timely involvement in economic turnover.

Improving environmental culture and education of the population.

The President's message pays great attention to the issue of ensuring environmental protection and environmental safety in accordance with international standards [107].

In 2006, the Environmental Code was adopted, aimed at harmonisation of the environmental legislation of the Republic of Kazakhstan with advanced international acts, transition to new standards, improvement of the state control system [108]. To fulfil this goal, our generation needs environmental knowledge, based on which the future generation of Kazakhstanis will be brought up. For this purpose, it is necessary to conduct environmental education of students in secondary schools in close connection with career guidance [109].

At present, young people have a somewhat one-sided view of ecology. They believe that ecology is the science of the environment and the relationship of living organisms, but for some reason they often do not include themselves in the system of these relations.

It should be recalled that ecology (from Greek "oikos" - house, dwelling and "logos" - teaching) is a science that studies the conditions of existence of living organisms and the relationship between organisms and the environment in which they live [110]. The subject of ecology is the totality or structure of relationships between organisms and the environment. The main object of study in ecology is ecosystems, i.e. unified natural complexes formed by living organisms and their environment [111]. In addition, its field of competence includes the study of individual species of organisms, their populations and the biosphere as a whole.

The main traditional part of ecology as a science is general ecology, which studies the general patterns of relationships between any living organisms and the environment, including humans [112].

All-round development of personality, formation of aesthetic, ecological, moral and creative elements of spiritual culture in schoolchildren is one of the tasks of vocational guidance work at school. And the task of teachers is to educate future young workers and specialists so that when they leave school, they would be able to bring beauty into life, into labour, into people's relations. The main elements of spiritual culture cannot be formed each separately [113]. They are closely interconnected with each other [114]. Speaking about ecology, human attitude to the environment, it is impossible not to mention the beauty of the animate and inanimate world, the honour and duty of people to nature. The process of formation and development of ecological culture becomes an impulse of spiritual and practical activity aimed at overcoming the crisis state of the system "society-nature", at improving this state, and in the long term - at harmonising relations between society and nature. "Charge" for the improvement of nature management should become a common feature of all social groups and generations, but especially of young people [115]. The generation entering independent life is the most receptive to new principles and norms of relations with nature, is characterised by the spirit of innovation, energy and other qualities so necessary for the implementation of these principles and norms. This need encourages new and new people to understand the environmental situation, to understand the causes of its unfavourable conditions, to comprehend the ways to improve it, and thus "leads" to environmental knowledge, stimulates its mastering [110].

In order to form environmental culture and educate the population, it is proposed to carry out the following activities:

- carrying out activities to inform the population of the region about the ecological situation and environmental protection through mass media and propaganda of ecological knowledge, including: production of visual agitation, publication of articles in mass media, production and broadcasting on TV of commercials on environmental topics, publication of information and environmental bulletins, subscription to periodical printing of ecological content of educational institutions of the region;

- Organisational and mass events for the environmental education of children and young people, such as competitions, meetings and conferences;

-          creation of a unified methodological centre for environmental education;

- Strengthening the ecological and world outlook of education, and, above all, broader coverage of philosophical problems of interaction between man and nature;

- pairing various school subjects with environmental issues, formation of interdisciplinary links identified in the course of development of interdisciplinary studies of nature protection and environmental rehabilitation;

- development and introduction of holistic training courses in nature protection and general ecology, which reflect in the educational process such a trend of scientific knowledge as the formation and development of holistic integrated areas of environmental research;

- inclusion in environmental education of the results of those scientific researches in the field of environment, which are related to regional and sectoral specialisation.

Environmental expertise, forecasting and monitoring.

The solution of environmental problems is impossible without scientific environmental expertise and forecasting. Environmental expertise is an assessment of the environmental situation in the area of operation of an existing or projected enterprise. Based on the analysis of the current situation (for the designed enterprises - by the method of analogues), priority measures to reduce the negative impact on the environment are developed [95].

Forecasting - prediction of the situation development, development of scientifically grounded judgements about the state of the natural environment and its development trends for decision-making on rational nature management. In terms of time coverage the forecast can be short-term, medium-term and long-term. The forecast can be regional (for the whole region or a large part of it) and local - in the sphere of impact on the natural environment of a particular large enterprise.

The forecast can be exploratory, answering the question of what is most likely to happen if the existing trends are maintained, or normative, showing how to achieve the desired states on the basis of given norms, goals. Such a forecast answers the question of how, by what ways to achieve the desired [98].

Forecasting is carried out on the basis of special eco-geographical studies using methods of analogy, expert assessments, statistical extrapolation and simulation modelling. It is extremely important to develop comprehensive forecasts that cover both nature and society, forecasts of the dynamics of geosystems, ecological and economic systems, which should link the maximum possible rates of economic growth, social development and environmentally acceptable level of pollution (or quality) of the natural environment [110].

Monitoring is an information system, the main tasks of which are observation and assessment of the state of the natural environment affected by anthropogenic impact for the purpose of rational use of natural resources and environmental protection. The most developed monitoring system is the control of water and air pollution.

As a result of implementation of the proposed measures to improve environmental problems of Akmola region, conditions for stabilisation and improvement of environmental quality will be created and the following results should be achieved:

- reducing the level of anthropogenic impact on the environment;

- reduction of air pollutant emissions from mobile sources when installing gas-cooled equipment on motor vehicles (reduction of carbon monoxide, nitrogen oxides, hydrocarbon emissions, no emissions of environmental dioxide and lead compounds, reduced smokiness of exhaust gases);

- measures to create sanitary protection zones of green areas on busy motorways of the city and the region will reduce the concentration of harmful emissions into the atmospheric air from mobile sources;

- carrying out measures to improve water quality by introducing additional treatment systems taking into account groundwater research, designing local water supply facilities;

- reconstruction and overhaul of sewerage facilities will help avoid emergencies and exclude the entry of pollutants into rivers and lakes;

- the system of waste storage, disposal and recycling is being improved, the waste management programme will allow solving the problems of waste management in the region; construction of a toxic waste landfill, reconstruction of a landfill will allow avoiding unfavourable environmental situations;

- the level of population morbidity associated with unfavourable surface water, atmospheric air and radiation conditions will decrease;

- The system of environmental education, upbringing and enlightenment is being improved.

Implementation of the environmental management system 180 14001:2004 at the enterprises of the region.

The 180 14000 standards are "voluntary". They do not replace legal requirements, but provide a system for determining how a company affects the environment and how legal requirements are met [115].

An organisation may use the 180 14000 standards for internal use, for example as an EM8 model or an internal audit format for an environmental management system. It is intended that the establishment of such a system provides an effective tool for an organisation to manage the totality of its environmental impacts and to bring its activities into compliance with a variety of requirements.

Standards can also be used externally to demonstrate to customers and the public that the environmental management system is up-to-date. Finally, an organisation may seek formal certification from a third (independent) party. As can be surmised from the experience of the 180 9000 standards, it is the desire to obtain formal registration that is likely to drive the adoption of environmental management systems that comply with the standard. Despite the voluntary nature of the standards, according to Jim Dixon, chairman of I80/TC 207 (the technical commission developing 180), in 10 years from 90 to 100 per cent of large companies, including multinational companies, will be certified to 180 14000, i.e. will receive a "third party" certification that certain aspects of their activities comply with these standards [112].

Businesses may wish to obtain 180 14000 certification primarily because such certification (or registration in 180 terminology) will be one of the prerequisites for marketing products on international markets (for example, the EEC recently announced its intention to allow only 180-certified companies to market in Commonwealth countries).

Other reasons why a business may need EM8 certification or implementation include:

- improving the company's image in the area of environmental compliance (including environmental legislation);

- saving energy and resources, including those allocated to environmental protection measures, through more efficient management;

- increase in the estimated value of the company's fixed assets; desire to conquer markets for "green" products;

-        improvement of the enterprise management system;

-        interest in attracting highly skilled labour.

By design, the certification system should be established at the national level. Judging by the experience of such countries as Canada, the leading role in the process of creating a national certification infrastructure is played by national standardisation agencies, such as Gosstandart, as well as chambers of commerce and industry, business unions, etc.

The standard registration process is expected to take between 12 and 18 months, about the same amount of time it takes to implement an environmental management system at a company.

Since the requirements of 150 14000 overlap in many ways with 150 9000, a lighter certification is possible for companies that already have 180 9000. In the future, it is envisaged that "dual" certification will be possible to reduce the overall cost [113].

The following recommendations can be given as specific recommendations for each surveyed district of Akmola region:

Akkol district.

The analysis of the existing situation of atmospheric air pollution shows that there is a significant pollution of atmospheric air in settlements. In order to reduce concentrations of pollutants in the atmospheric air it is necessary to intensify work in the following directions:

1.   Installation of dust and gas cleaning equipment at industrial enterprises and implementation of measures to reduce emissions established by the conclusions of the state environmental expertise;

2.   Development and implementation of urban planning measures aimed at reducing exhaust gas concentrations in human areas.

The analysis of wastewater discharge dynamics shows that there is a slight annual increase, which is mainly due to the increase in water consumption. The lack of organisational wastewater reception in rural settlements has a negative impact on the environment by polluting ground and surface waters and land and biological resources.

In order to solve the issues of water disposal in the settlements of the district it is necessary to solve the following tasks:

1.   In the district centre of Akkol city to provide for reconstruction of sewerage collector. Develop a project of maximum permissible wastewater discharges to filtration fields, which will allow to assess the need for construction of treatment facilities;

2.   In rural settlements it is necessary to organise wastewater removal to specially equipped sites for its reception.

Analysis of the dynamics of waste generation shows that there is a rather high increase compared to 2012. The main reasons for the increase in waste generation is the removal of rubbish during the sanitary three-month period, the formation of which occurred in previous years and a sharp increase in construction waste, the formation of which occurs from the liquidation of demolished buildings and structures.

The main areas of waste management regulation are:

1.   Development of settlement dumps for waste disposal in accordance with the existing regulations;

2.   Establishment of enterprises for waste removal and routine works at landfills and dumps;

3.   To envisage allocation of budgetary funds for implementation of the programme "Waste management of Akmola region" according to the measures

proposed in the plan of this programme.

4.   Strengthen control and environmental propaganda to prevent the emergence of unauthorised landfills.

Zerenda district.
To reduce concentrations of pollutants in the atmospheric air, it is necessary to intensify work in the following areas:

- Installation of dust and gas cleaning equipment at industrial enterprises and implementation of emission-reducing measures established by the conclusions of the state environmental expertise.

To solve the issues of centralised wastewater disposal from enterprises and settlements of the district, it is necessary to implement the following measures:

1. Development of a project of maximum permissible discharges from the village of Zerenda. Zerenda village, which will allow to exclude groundwater pollution and obtain justification of the necessity to construct treatment facilities;

2. In settlements where there is no sewerage system, it is necessary to organise the reception and removal of wastewater to specially equipped sites for its reception.

The main focus on addressing solid waste management issues is:

1. Establishment of enterprises for waste removal and routine works at landfills and dumps;

2. Strengthen control and environmental propaganda to prevent the emergence of unauthorised landfills.

Burabai district.

The analysis of the existing situation of atmospheric air pollution shows that there is a significant pollution of atmospheric air in settlements. In order to reduce concentrations of pollutants in the atmospheric air it is necessary to intensify work in the following directions:

1. Installation of dust and gas purification equipment at industrial enterprises and implementation of emission-reducing measures established by conclusions of the state environmental expertise;

2.   Development and implementation of urban planning measures aimed at

reducing exhaust gas concentrations in human areas.

3. In the future - heat supply of Shchuchinsk and other settlements of the district from a single source (a single source in each settlement) operating on the most environmentally friendly fuel - gas. For this purpose it will be necessary to conduct a gas pipeline through the territory of the rayon.

4. Increase the frequency of the network of motor transport control posts and tighten their work in order to allow motor vehicles to enter the resort area on a portion basis.

To solve the issues of wastewater disposal in Burabai district it is necessary:

1. Complete the construction of a biological wastewater treatment plant at the Shchuchinsko-Borovsky resort area.

2. Put into operation in Burabai district Kokshetau industrial water pipeline to reduce water intake from Shchuchye lake and underground wells.

The main areas of waste management regulation are:

1. Development of settlement dumps for waste disposal in accordance with the existing regulations;

2. Establishment of enterprises for waste removal and routine works at landfills and dumps;

3. Strengthen control and environmental propaganda to prevent the emergence of unauthorised landfills.

4. Place as many rubbish collection containers as possible in the resort area with regular removal to landfills.

5. Establish special brigades to collect rubbish in the resort area at the "Burabai" State Research and Production Enterprise and regularly conduct by these brigades relevant   raids   in

forests.

CHAPTER 4

CONCLUSION

Traditionally, the state of health of the population is characterised by a system of statistical indicators that determine the features of population reproduction (medical and demographic characteristics), the stock of physical strength and capacity (indicators of physical development of the population), the features of adaptation of the population to environmental conditions (morbidity of the population) [74].

The level of population health is a summary resultant indicator of interaction between individuals and the environment. In modern conditions, the active relationship between humans and the environment leads to significant changes and complication of ecology, which necessitates research aimed at a thorough study of population health, the search for effective criteria of its state for monitoring and predicting changes [54].

Evaluation of the completeness of the solution of the set tasks. The aim of the work has been achieved and the research tasks have been fully solved, which confirms the reliability of the main conclusions and provisions of the thesis:

- a comprehensive assessment of environmental pollution in some districts of Akmola region on ecological and hygienic indicators of atmospheric air, drinking water and soil was carried out;

- The influence of a complex of environmental factors on the peculiarities of the formation of health of children, adolescents and adults in terms of morbidity has been revealed;

- The causal relationship between environmental factors and health indicators has been established;

- recommendations are given to reduce the impact of technogenic load on the health of the population.

The study results in the following conclusions and findings:

1. The districts of the region have their own specifics of air pollution, depending on the nature of industries located there.

The most environmentally unfavourable among the three studied districts of

Akmola region is Akkol district, here exceeding the average daily concentrations of formaldehyde, sulphur dioxide and carbon monoxide is observed. In Burabay district the situation is the most favourable, no exceedance of maximum permissible level is observed for any of the substances considered. In Zerendinsky district there is a small excess of MPC s.s. for sulphur dioxide, in general, the environmental situation is favourable.

2.	Water supply in farms of Burabay rayon is extremely unsatisfactory. Water distribution networks are not washed in time and disinfected. Water pipelines in rural areas are in poor sanitary and technical condition. Villagers are forced to use water from open water bodies and dig wells.

3.	Chemical analysis of soil of the study areas showed that the content of such substances as cadmium, zinc, lead, chromium do not exceed MPC. However, 1.5 times of copper exceeds MAC in the soil of Akkol district and 0.8 times in the soil of Zerendinsky district.

4. It was found that children and adolescents living in the regions of the region with different levels of atmospheric air pollution have no differences in physical development indicators, but normal physical development is observed only in 52.6 per cent of the subjects.

5. It has been established that in the areas where CHPPs are located, the incidence of respiratory diseases, circulatory system, blood diseases and haematopoiesis is 1.5 - 2 times higher among children, adolescents and adults.

6. A direct correlation between the indicators of atmospheric air pollution and blood diseases, congenital anomalies of development, diseases of the endocrine system, respiratory organs, neoplasms, diseases of the nervous system and sensory organs has been established ($r = 0.44 - 0.73$).

Recommendations on how to utilise the results of the work.

In the course of the study the territories of Akmola region with different levels of environmental ill-being, priority pollutants and their impact on health were identified, which makes it possible to use the obtained data in the development of preventive measures aimed at reducing the impact of anthropotechnogenic load on the health of the population.

LIST OF SOURCES USED

[1] Nazarbayev N.A. Increase of welfare of citizens of Kazakhstan - the main goal of the state policy: the message of the President of the Republic of Kazakhstan to the people of Kazakhstan. - Astana, 2008.- P. 55-57.

[2] Imambayeva T.M. Clinic and treatment of asthmatic status in children living in areas of ecological disadvantage // Collection of AGMI: "Problems of ecology in pathophysiology".-Almaty, 1995.-P. 157- 164.

[3] Makhanov T.M., Saduakasova A.S., Tuleutaev K.T. Health of the population living in the zone of ecological disadvantage. // scientific and practical conference on topical issues of practical medicine.- Almaty - Kyzylorda, 1996.- P. 12-14. 12-14.

[4] Zhakashov N.J. Methodological and social aspects of mortality of the population of cities with different intensity of environmental pollution// Issues of environmental hygiene.-Almaty,1992.- P.122-129

[5] Sabirova Z.F. Anthropogenic pollution of atmospheric air and health status of children's population // Hygiene and Sanitation,- 2001.- No. 2, - P. 7-9. 7-9.

[6] Stepanova N.V. Immune status of children in conditions of pollution of a large city by heavy metals // Hygiene and Sanitation. - 2004.- №5. -C.42-44.

[7] Serdyukovskaya G.N. Influence of environmental factors on the health of the younger generation // Bulletin of the Academy of Medical Sciences. SSSR.- 1986.- №3.- PP.135-137.

[8] Godina G.Z., Miklashevskaya N.N. Some trends in somatic development of urban children and adolescents over the past 20 years: on the example of examination of schoolchildren in Moscow // Bulletin of the USSR Academy of Medical Sciences.- 1990.- №8- 79p.

[9] Mazhibaev K.A., Ormantaev K.S., Khusainova Sh.N. State of health of children of Kazakhstan and prospects for the development of paediatric service and science. // Materials of I(V) Congress of Children's Doctors of the Republic

of Kazakhstan.- Astana, F2001.-C.18-19.

[10] Sarycheva *S.Y.,* Apatov V.V., Grebnyak V.I. Atmospheric pollution as one of the socio-hygienic factors determining the state of health of schoolchildren // Health protection of children and adolescents. Zdorovye.- 1984.-Vyp. 15.- C. 17-20.

[11] Khabizhanov B.H., Ishuova P.K., Kumusbaeva R.S., et al. Characteristics of vascular dystonia and functional changes of the heart in children of the Aral Sea region. // Materials of the conference "Problems of ecological medicine", - 1993, -Ch.2.-C.94-96.

[12] Umansky V.Ya. Hygienic bases of assessment of early violations of children's health under the influence of unfavourable environmental factors on the health of the younger generation // Bulletin of the USSR Academy of Medical Sciences.- 1981.- № 3.- P. 135- 137.

[13] Belyakov V.A., Vasiliev A.V. Influence of atmospheric air pollution on physical development of children // Hygiene and Sanitation.- 2003.- №4.- P.49- 51.

[14] Sultankulova J.V., Sarsenbaeva S.S. State of the urinary system in children as an indicator of environmental situations. // Materials of the I(V) Congress of Children's Doctors of the Republic of Kazakhstan.- Astana, 2001.- P. 186. 186.

[15] Selivanov A.P., Yampolskaya I.Y., Tomash V.V.. Atmospheric pollution as one of the social-hygienic factors determining the state of health of schoolchildren // Health protection of children and adolescents.- 1984. Issue. 15.- C. 17-20.Krivtsova S.V. et al. Teenager at the crossroads of epochs. M.,: Genesis. 1997, 288 pp.

[16] Koshkina E.A. Methodical approaches to the development of a comprehensive assessment of the state of health of children of early and preschool age: Actual issues of social hygiene and health care organisation // Zdravookhranenie Ross. Feder,- 1975.- №5.-S.14-16.

[17] Chikisheva T.A. Studying the links between anthropological features and environmental factors (on the example of the Altai-Sayan region). Novosibirsk,

1992.- 162 p.

[18] Roberts D. Body weiyth, race and climate // Amer. J. Phius. Antropol. - 1983. - Vol. 2. 4. - P. 533-558.

[19] Kurmanalin B.A., Zhumalina A.K., Pukhovikova H.H. Physical development of school-age children living near a gas processing plant // Paediatrics and children's surgery. -2003.- №4. -С. 8-9.

[20] Burce E. Measures of body composition and perforamance in maior collage football player // J. Sports. Sports. Med. Fhys. Fithess. - 1980. - № 20. - P. 176-180.

[21] Mtridish H.V. Relation between socioleconomik status and bab size seben to ten yeans age // Amer. J. Dis. Chiolog. - 1991. - Vol. 82. - P. 702-709.

[22] Petrov V.G. Medico-social aspects of health of the population of the regions of ecological disaster of Kazakhstan // Materials of the scientific conference devoted to the 50th anniversary of formation of the Institute of Research Institute of Hygiene and Disease Prevention. -Almaty, 1994.- 226 pp.

[23] Borisov B.M., Primakov V.I., Martirova T.A. Ecological approaches in assessing the state of health of adolescents // Military - Medical Journal.- 1996.- № 2.-S. 51- 52.

[24] Agaev F.B., Samedov I.G., Kuliev A.S. Quantitative and qualitative assessment of the relationship between the morbidity of infants with chemical pollution of the atmosphere in the conditions of Baku // Hygiene and Sanitation. - 1993. - №4. - C.76.

[25] Belyaev E.N. Role of sanitary-epidemiological service in ensuring sanitary-epidemiological well-being of the population of the Russian Federation. - M., 1996.-416 p.

[26] Boev V.M. Hygienic characterisation of the influence of anthropogenic and natural geochemical factors on the health of the population of the Southern Urals // Hygiene and Sanitation. - 1998. - №6. - C.3-8.

[27] Water resources of Kazakhstan in the new millennium. A. 2004г.-132с.

[28] A. Kazbekov. The most precious fossil of the earth, Kokshetau.2002.-272

p.

[29] Ecological Information Bulletin.- Almaty: REC, 20122013.

[30] Bolshakov A.M., Dmitriev A.D. Contribution of environmental factors in the peculiarities of ontogenetic processes // Hygiene and Sanitation. - 1993. - №6.- C.75-77.

[31] Bukharin O.V., Litvin V.Yu. Pathogenic bacteria in natural ecosystems. - Ekaterinburg: Ural Branch of the Russian Academy of Sciences, 1997. - 277 c.

[32] Bushtueva K.A., Sluchanko I.S. Methods and criteria for assessing the state of public health in connection with environmental pollution. - Moscow: Medicine, 1979. - 160 c.

[33] Bystrykh V.V. Complex hygienic assessment of environmental pollution of an industrial city and health indicators of newborns. Avtoref. diss. .kand. med. sciences. - Orenburg, 1995. - 23 c. [34] Bystrykh V.V., Boev V.M. Atmospheric pollution and anthropometric indicators of newborns in Orenburg // Hygiene and Sanitation. - 1995. - №1. - C.3-4.

[35] Bystrykh V.V., Boev V.M., Borshchuk E.L. Assessment of additional carcinogenic risk in connection with anthropogenic pollution of atmospheric air of residential areas // Hygiene and Sanitation. - 1999. - №1.- C.8-10.

[36] Bystrykh V.V., Boev V.M., Borshchuk E.L. et al. Using the methodology of estimation of hydrogen sulfide toxic impact hazard // Problems of prevention and liquidation of emergency situations consequences on pipelines of oil and gas complex: Proc. of All-Russian scientific-practical conf. - Orenburg, 1998. - C.114.

[37] Bystrykh V.V., Boev V.M., Borshchuk E.L., Dunayev V.N. Air pollution in the area of motorway as a risk factor // Ecology of big city: Proc. of scientific and practical conference - Perm, 1996. - C.14-15.

[38] Bystrykh V.V., Boev V.M., Borshchuk E.L., Kudrin V.I. Assessment of additional carcinogenic risk in an industrial city // Environment. Health Risk Assessment. Experience of application of risk assessment methodology in Russia. - M., 1998. - Vyp.5. - P.22-23.

[39] Veltischev Yu.E. Problems of ecopathology of childhood - immunological aspects // Paediatrics. - 1991. - №12. - C.74-80.

[40] Veltischev Y.E., Fokeeva V.V. Ecology and children's health (ecotoxicological direction) // Maternity and Childhood. -1992. - №12.- C.30-35.

[41] Detiuk E.S., Datsenko I.I., Avgustinovich M.S. et al. Influence of atmospheric air pollution on morphofunctional indices of the placenta // Hygiene and Sanitation. - 1991. - №6. - C.10-12.

[42] Zaitseva N.V., Averyanova N.I., Koryukina I.P. Ecology and health of children of the Perm region. - Perm, 1997. - 147 c.

[43] Ivanov V.Y., Tokarev I.I., Kulikova T.E. Population morbidity associated with atmospheric air pollution in Zaporozhye // Hygiene and Sanitation. - 1993. - №6.- C. 11-13.

[44] Kiryushchenkov A.P., Tarakhovsky M.L. Effect of medicinal substances on the foetus. - M., 1990. - 272 c.

[45] Knizhnikov VA, Novikova KV, Grozovskaya VA et al. To the question of blastomogenic efficiency of the combined action of components of fly coal ash // Hygiene and Sanitation. - 1987. - №3. - C.10-13.

[46] Koskina EV, Bonashevskaya TI, Barkov LV System of indicators fetoplacental complex to assess the state of atmospheric air // Hygiene and Sanitation. - 1992. - №2. - C.14-17.

[47] Kryatov I.A., Avkhimenko M.M., Tsapkova N.N. Polychlorinated biphenyls and dioxins - dangerous and persistent environmental pollutants (review)//Hygiene and Sanitation-1991- № 12. - C.68-72.

[48] Kuznetsova T.P. State of health of newborns in female workers of oil refinery // Zdravookhranenie Rossiyskoy Federatsii. - 1992. - №6. - C.27-28.

[49] Kutepov E.N. Methodical bases of estimation of the state of health of the population under the influence of environmental factors. Avtoref. diss. of doctor of medical sciences. - M., 1995. - 41 c.

[50] Kuchma V.R., Hildenskiold S.R., Minibaev T.Sh. et al. Epidemiology of

diseases of the population living in ecologically unfavourable territories // Ecological safety of regions and market relations: Materials of the international conference. - M., 1994. - C.363-368.

[51] Lebedkova S.E., Boev V.M., Kolbina L.V. et al. Prevalence of cardiovascular diseases in school-age children's population taking into account the ecological situation of the air environment //Pediatrics. - 1991. - №12. - C.41^4.

[52] Likhachev A.Ya. Study of environmental pollution by carcinogenic substances and the possibility of predicting individual sensitivity to them // Voprosy Onkologii. - 1997. - №1.- C.111-115.

[53] Muzaleva O.V. Complex hygienic assessment of anthropogenic pollution and characterisation of staphylococcal autoflora in schoolchildren of an industrial city. Auth. diss. ... Cand. of medical sciences. - Orenburg, 1999. - 26 c.

[54] Naumenko O.A. Epidemiology and monitoring of risk factors of diseases of cardiovascular system in schoolchildren living in conditions of a large industrial city. Avtoref. dis. ... Cand. med. sciences. - Orenburg, 1996.

[55] Nesterenko SA, Lineva OI Social ecology and its impact on immune homeostasis during pregnancy // Ecology and human health: Abstracts of the All-Russian scientific and practical conference. - Samara, 1994. - C. 120-122.

[56] Novikov S.M., Rumyantsev G.I., Zholdakova Z.I. et al. The problem of carcinogenic risk assessment of chemical environmental pollution // Hygiene and Sanitation. - 1998. - №1. - C.29-34.

[57] Parfenov Y.D. Calculation of maximum permissible concentration of beryllium in the air according to the criterion of carcinogenic effect // Hygiene and Sanitation. - 1988. - №6. - C.59-62.

[58] Pushkareva M.V. Criteria and methods of minimising the impact of environmental loads on the population. Auth. diss. ... Doctor of Medical Sciences. - Moscow, 1995.-44 p.

[59] Senichenkova I.N. About embryotoxic effect of industrial environment

pollutants - formaldehyde and petrol // Hygiene and Sanitation. - 1991. - №9. - C.35-38.

[60] Sidorenko G.I., Kutepov E.N.. Priority directions of scientific research on the problems of assessment and forecasting the impact of risk factors on the health of the population // Hygiene and Sanitation. - 1994. - №8. - C.3-5.

[61] Electronic resource, http://referat.resurs.kz/

[62] Electronic resource, http://www.ref.by/

[63] Electronic resource, http://ecounion.ecorussia.info/ru/

[64] Electronic resource. http://library.ipae.uran.ru/

[65] Electronic resource, http://ecology.uran.ru/

[66] Slivina L.P., Popov S.V., Voronkova O.A. et al. Risk factors of diseases of children of the first year of life in a large industrial city //Actual problems of hygiene: Proc. of scientific conf. - Kazan, 1994. - C.6769.

[67] Sokolov V.V., Frasch V.N. Discussion issues of leukemogenic (blastomogenic) action of benzene //Hygiene of labour and occupational diseases. - 1985. - №4. - C.21-26.

[68] Sychev A.A., Sannikov V.M. Integrated methodological approach to the assessment of genetic consequences of atmospheric air pollution //Hygiene of the environment. - Kiev, 1989. - C. 149-150.

[69] Usvyatsov B.Y., Muzaleva O.V., Gerbich I.I. et al. Hygienic evaluation of staphylococcal biocenosis of the nasal mucosa of schoolchildren in an industrial city // Hygiene and Sanitation. - 1998. - №6. - C.13-16.

[70] Fateeva T.A., Setko N.P., Stadnikov A.A. Experimental studies of the effect of multi-sulphur natural gas and condensate on reproductive function // Hygiene and Sanitation. - 1998. - №5. - C.5-7.

[71] Filov V.A., Khudoley V.V. Chemical carcinogens in the environment and their ecological significance. Natural and anthropogenic carcinogens // Journal of Ecological Chemistry. - 1993. - №4. - C.313-317.

[72] Aggett P.J., Rose S. Soil and congenital malformations // Experientia. - 1987. - Vol.43, No.1. - P.104-108.

[73] Aksoy M. Hematotoxicity and carcinogenicity of benzene // Environ. Health. Perspect. - 1989. - Vol.82. - P.193-197.

[74] Ayotte P., Livesque B., Gauvin D. et al. Indoor exposure to 222Rn: a public health perspective // Health Phys. - 1998. - Vol.75, No.3. - P.297-302.

[75] Baker F.D., Bush B., Tumasonis C.F. et al. Toxicity and persistence of low-level PSB in adult Wistar rats, fetuses and young // Arch. Environ. Contam. and Toxicol. - 1977. - Vol.5, No.2. - P.143-156.

[76] Bako G., Smith E.S., Hanson J., Dewar R. The geographical distribution of high cadmium concentrations in the environment and prostate cancer in Alberta // Can. J. Public. Health. - 1982. - Vol.73, No.2. - P.92-94.

[77] Blair A., Kazerouni N. Reactive chemicals and cancer // Cancer Causes Control. - 1997. - Vol.8, No.3. - P.473-490.

[78] Boezen H.M., van der Zee S.C., Postma D.S. et al. Effects of ambient air pollution on upper and lower respiratory symptoms and peak expiratory flow in children // Lancet. - 1999. - Vol.353. - P.874-878.

[79] Borchers M.T., Carty M.P., Leikauf G.D. Regulation of human airway mucins by acrolein and inflammatory mediators // Am. J. Physiol. - 1999. - Vol.276. - P.L549-L555.

[80] Csicsaky M.J., Roller M., Pott F. Risk modelling: which models to choose? // Exp. Pathol. - 1989. - Vol.37, №1-4. - P.198-204.

[81] Dypbukt J.M., Atzori L., Edman C.C., Grafstrom R.C. Thiol status and cytopathological effects of acrolein in normal and xeroderma pigmentosum skin fibroblasts // Carcinogenesis. - 1993. - Vol.14, No.5. - P.975-980.

[82] Ekman P. Genetic and environmental factors in prostate cancer genesis: Identifying high-risk cohorts // Eur. Urol. - 1999. - Vol.35, No.5-6. - P.362369.

[83] Kipling M.D. Oil and cancer // Ann. R. Coll. Surg. Engl. - 1974. - Vol.55, No.2.- P.71-79.

[84] Klein C.B., Kargacin B., Su L. et al. Metal mutagenesis in transgenic Chinese hamster cell lines // Environ. Health. Perspect. - 1994. - Vol.102. - Suppl.3. - P.63-67.

[85] Lagarde F., Pershagen G. Parallel analyses of individual and ecological data on residential radon, cofactors, and lung cancer in Sweden // Am. J. Epidemiol. - 1999. - Vol. 149, №3. - P.268-274.

[86] Kalitsun V.I., Laskov Y.M. Laboratory workshop on water disposal and wastewater treatment. M. Stroyizdat - 1995. - C. 325.

[87] Shiklomanov I.A. Land water resources research. L.: Gidrometeoizdat, 1988. - 152 c.

[88] Fomin G.S. Water. Control of chemical, bacterial and radiation safety according to international standards. Encyclopaedic reference book. M. 2000-848 p.

[89] Practicum on ecology and environmental protection. A.I. Fedorova, A.N. Nikolskaya. Moscow: Vlados, 2001. - 288 c.

[90] Kaurichev I.S. Practicum on soil science. Moscow, 1986. - 425 c.

[91] Toth E., Ulveczky E., Verebelyi Z. Oroszlan G. Lead concentration in human milk in high aerosured population // Eur. Congr. Perinatal Med. -Roma, 1989. - Vol.2. - P.1097-1099.

[92] Winneke H., Klingenberg H. Studies on health effects of automotive exhaust emissions. How dangerous are diesel emissions? // Sci. Total Environ. - 1990. - Vol.93. - P.95-105.

[93] Yang C.S. Research on esophageal cancer in China: a review //Cancer. Res. - 1980. - Vol.40. - P.2633-2644.

[94] Ospanova G.S., Bozmataeva G.T. Ecology. - Almaty: Ekonomika, 2002. - 405

[95] Electronic resource, http://referat.resurs.kz/

[96] Electronic resource, http://www.ref.by/

[97] Electronic resource, http://ecounion.ecorussia.info/ru/

[98] Electronic resource, http://library.ipae.uran.ru/

[99] Electronic resource, http://ecology.uran.ru/

[100] Environmental Code of the Republic of Kazakhstan. - Almaty: YURIST, 2013. - 300

[101] Methodological Recommendations to Sanitary Rules and Norms for Protection of Surface Water from Pollution (SanPiN No. 3.02.003-04). http://doword.ru/

[102] Methodology for calculation of maximum permissible discharges (MPD) of substances to the

[103] Instruction on rationing of pollutant discharges into water bodies (RND 211.2.03.01-97). - Almaty: RARITET, 1997. - 190 c.

[104] Electronic resource, http://masters.donntu.edu.ua/

[105] Sorokina N.D. Environmental protection at the enterprise. - Moscow: Khimiya, 2014. - 200 c.

[106] Rules for Protection of Surface Waters of the Republic of Kazakhstan (RID 1.01.03-94). - Almaty: JURIST, 1994. - 222 c.

[107] Methodical instructions on application of "Rules of protection of surface waters of the Republic of Kazakhstan", introduced on 01.07.94 (RID 11.2.03.02-97). - Almaty: JURIST, 1997. - 200 c.

[108] Assignments and methodological instructions for practical classes on the course Ecology, Nature Management, http:// bib.convdocs.org/v20410/

[109] ESCO. Cities and Buildings, http://www.journal.esco.co.ua/

[110] Eltermap V.M. Environmental protection. - Moscow: Khimiya, 1995. - 360 c.

[111] Digital Library, http://www.himi.oglib.ru/

[112] Membrane treatment of waste water from industrial enterprises. http://otherreferats. allbest.ru/ecology/.

[113] Medrim G.L., Teisheva A.A., Basin D.A. Disinfection of natural and waste water using electrolysis. - Moscow: Stroyizdat, 1999. - 460 c.

[114] Technical Notes on Water Quality Problems: Per.s Engl. Edited by Karyukhina T.A., Churbanova I.N. - M.: Nauka, 2000. - 608 c.

[115] Loginov O.N. Biotechnological methods of environmental purification from anthropogenic pollution / O.N. Loginov, N.N. Silischev, T.F. Boyko, N.F. Galimzyanova. - Ufa: State Publishing House of Scientific and Technical

Literature "Reaktiv", 2000. - 100 c.

Printed by Books on Demand GmbH, Norderstedt / Germany